Springer Theses

Recognizing Outstanding Ph.D. Research

Aims and Scope

The series "Springer Theses" brings together a selection of the very best Ph.D. theses from around the world and across the physical sciences. Nominated and endorsed by two recognized specialists, each published volume has been selected for its scientific excellence and the high impact of its contents for the pertinent field of research. For greater accessibility to non-specialists, the published versions include an extended introduction, as well as a foreword by the student's supervisor explaining the special relevance of the work for the field. As a whole, the series will provide a valuable resource both for newcomers to the research fields described, and for other scientists seeking detailed background information on special questions. Finally, it provides an accredited documentation of the valuable contributions made by today's younger generation of scientists.

Theses may be nominated for publication in this series by heads of department at internationally leading universities or institutes and should fulfill all of the following criteria

- They must be written in good English.
- The topic should fall within the confines of Chemistry, Physics, Earth Sciences, Engineering and related interdisciplinary fields such as Materials, Nanoscience, Chemical Engineering, Complex Systems and Biophysics.
- The work reported in the thesis must represent a significant scientific advance.
- If the thesis includes previously published material, permission to reproduce this must be gained from the respective copyright holder (a maximum 30% of the thesis should be a verbatim reproduction from the author's previous publications).
- They must have been examined and passed during the 12 months prior to nomination.
- Each thesis should include a foreword by the supervisor outlining the significance of its content.
- The theses should have a clearly defined structure including an introduction accessible to new PhD students and scientists not expert in the relevant field.

Indexed by zbMATH.

Takahiro Doba

Iron-Catalyzed C–H/C–H Coupling for Synthesis of Functional Small Molecules and Polymers

Doctoral Thesis accepted by
The University of Tokyo, Tokyo, Japan

Springer

Author
Dr. Takahiro Doba
Department of Chemistry
The University of Tokyo
Tokyo, Japan

Supervisor
Prof. Eiichi Nakamura
The President's Office and the Department of Chemistry
The University of Tokyo
Tokyo, Japan

ISSN 2190-5053 ISSN 2190-5061 (electronic)
Springer Theses
ISBN 978-981-99-4120-9 ISBN 978-981-99-4121-6 (eBook)
https://doi.org/10.1007/978-981-99-4121-6

This Springer imprint is published by the registered company Springer Nature Singapore Pte Ltd.
The registered company address is: 152 Beach Road, #21-01/04 Gateway East, Singapore 189721, Singapore

Supervisor's Foreword

It is my great pleasure to have the opportunity to commend the doctoral thesis of Dr. Takahiro Doba, which has been published as a part of the esteemed Springer Theses. This comprehensive work highlights the exceptional research achievements of Takahiro during his Ph.D. study, which focused on the development of iron-catalyzed C–H/C–H coupling reactions that occur through iron-catalyzed C–H activation and their application to synthesizing functional small molecules and polymers.

The development of new methodologies and catalysts for catalytic C–H/C–H coupling reactions is an important goal in modern synthetic organic chemistry. Despite significant advances in the accessibility to complex organic materials and pharmaceuticals through established methods utilizing precious transition metals, there remains a strong need for new versions of C–H/C–H coupling utilizing readily available, environmentally friendly, and sustainable catalysts, particularly in consideration of the increasing attention to the Sustainable Development Goals (SDGs). The first part of Takahiro's thesis focuses on the development of iron-catalyzed C–H/C–H homocoupling of thiophene compounds using oxalate as a mild oxidant. The reaction is suitable for synthesizing electron-rich organic materials that are easily oxidized in the presence of strong oxidants, which is always required in the traditional C–H/C–H coupling due to the high redox potentials of precious transition-metal catalysts. Moreover, the efficiency of this homocoupling reaction was significantly improved by using newly designed tridentate phosphine ligands bearing a heteroaromatic group. With these ligands in hand, the polymerization of thiophene monomers was achieved, yielding functional polymers that can be used as hole-transporting materials for perovskite solar cells.

In addition, Takahiro developed the iron-catalyzed oxidative C–H alkenylation, or Fujiwara-Moritani reaction, by intercepting the homocoupling catalytic cycle with an enamine. This reaction proceeded through the nucleophilic addition of the enamine to the iron center, resulting in opposite reactivity compared to traditional reactions using palladium catalysts.

As a result of his groundbreaking research, Takahiro has received numerous honors, including Dean's Awards from the University of Tokyo in 2017, 2019, and 2022, the Student Presentation Award of the 101st CSJ Annual Meeting, Poster Prizes in domestic and international conferences in 2018 and 2021, Research Fellowship from the Japan Society for the Promotion of Science in 2020 and 2022, and the 12th Otsu Conference Award Fellow in 2021.

On behalf of the Department of Chemistry at the University of Tokyo, I extend my warmest congratulations to Takahiro for receiving the Springer Theses Award. I have no doubt that he will continue to explore new facets of chemistry and make significant contributions to the field, and I wish him all the best in his future endeavors.

May 2023

Prof. Eiichi Nakamura
University Professor
The President's Office
and the Department of Chemistry
The University of Tokyo
Tokyo, Japan

Abstract

Transition-metal-catalyzed $C(sp^2)$–H/$C(sp^2)$–H coupling has attracted much attention as one of the most straightforward methods to construct $C(sp^2)$–$C(sp^2)$ bonds. However, the application of this ideal transformation to the synthesis of redox-sensitive π-materials was hindered by the requirement of a strong oxidant for catalyst turnover. The purpose of this thesis is to demonstrate that iron is suitable for realizing transition-metal-catalyzed C–H/C–H coupling that is applicable to the synthesis of redox-sensitive functional small molecules and polymers. Thiophene compounds were chosen as substrates of interest because of the wide application found in materials science.

Chapter 1 describes the adequacy of using iron as a catalyst for the realization of transition-metal-catalyzed C–H/C–H coupling that operates under mildly oxidative conditions. Being aware that redox potential of catalyst is the limiting factor of the mildness of oxidant, the author focused on the low redox potential of Fe(III)/Fe(I) which has been partially proved by the use of dihaloalkanes as a mild oxidant in iron-catalyzed C–H activation reactions reported by Nakamura and co-workers.

Chapter 2 describes the development of iron-catalyzed regioselective thienyl C–H/C–H homocoupling using tridentate phosphine as a ligand, $AlMe_3$ as a base, and oxalate as a mild oxidant. Oxalate in combination with oxophilic $AlMe_3$ served as an effective but mild oxidant to oxidize Fe(I) species to regenerate Fe(III) catalyst. The electronic bias of the thienyl group helped to achieve regioselective C–H activation by a σ-bond metathesis mechanism.

Chapter 3 describes a modular synthesis of tridentate phosphine ligands by sequential addition of organolithium reagents to triphenyl phosphite. This method gave access to tridentate phosphine ligands having different substituents and accelerated further exploration of iron catalysis.

Chapter 4 describes the development of iron-catalyzed regioselective thienyl C–H/C–H polycondensation by improvement of iron-catalyzed regioselective thienyl C–H/C–H homocoupling introduced in Chap. 2. New tridentate phosphine ligands possessing a heteroaryl group were designed to suppress catalyst deactivation and employed as effective ligands for polycondensation. One of the obtained polymers was used as a hole-transporting material for perovskite solar cells.

Chapter 5 describes the development of iron-catalyzed oxidative C–H alkenylation of thiophene compounds with vinylcarbazoles and vinylindoles. The reaction proceeded with excellent regioselectivity, branched/linear selectivity, and *E*/*Z* selectivity to give a potential donor materials for optoelectronic device applications.

Finally, Chap.6 summarizes the present studies and gives further perspectives.

Acknowledgements

First, I wish to express my deepest gratitude to Professor, Dr. Eiichi Nakamura for his constructive advice, discussions, and constant encouragement throughout this work.
I would like to express my deep appreciation to a former Associate Professor, Dr. Laurean Ilies (now in RIKEN) and a current Associate Professor, Dr. Rui Shang for their valuable advice, discussions, and encouragement.

I am particularly grateful to Professor, Dr. Shū Kobayashi, Professor, Dr. Hiroyuki Isobe, Professor, Dr. Shuichi Hiraoka, and Professor, Dr. Jun Terao for insightful comments and suggestions about this thesis.

I am also grateful to Associate Professor, Dr. Koji Harano and Associate Professor, Dr. Takayuki Nakamuro for their advice, encouragement, and support throughout this work. I was especially impressed by Associate Professor, Dr. Takayuki Nakamuro's vitality to change the research field.

I am grateful to Professor, Dr. Shū Kobayashi and Assistant Professor, Dr. Tomohiro Yasukawa for valuable comments and suggestions on the removal of residual catalyst from polymers. I am also grateful to Professor, Dr. Yutaka Matsuo and Assistant Professor, Dr. Hao-Sheng Lin for the application of polymers to perovskite solar cells.

I am grateful to Professor, Dr. Klaus Müllen (MPIP) and Associate Professor, Dr. Akimitsu Narita (OIST) for my research experience in Germany from August to October in 2019.

I would like to show my appreciation to Dr. Tatsuaki Matsubara for his valuable comments and suggestions on my master course studies, iron-catalyzed C–H/C–H cross-coupling of carboxamides with heteroarenes. I would also like to show my appreciation to Dr. Wataru Sato for his technical comments and suggestions regarding polymer chemistry. My appreciation also goes to Mr. Mengqing Chen for the collaboration with the chromium-catalyzed C–H functionalization project. I am also thankful to Br. Shota Fukuma for his technical comments and suggestions regarding X-ray crystal structure analysis.

I would like to show my appreciation to Dr. Takumi Yoshida, Dr. Yuki Itabashi, Dr. Takenari Sato, Mr. Yi Zhou, Mr. Toki Go, Mr. Haotian Yang, and Mr. Naoki Matsushita for their valuable comments and suggestions on iron chemistry in a former subgroup. I would also like to show my appreciation to Dr. Hiroki Nishioka, Dr. Hiyoroshi Hamada, Mr. Takumi Sakamaki, Mr. Mengqing Chen, Mr. Olivier Chevalier, Mr. Mana Kawashima, Br. Shota Fukuma, Br. Akinori Takasugi, Br. Yosuke Miyazaki, and Shogo Aoki for their valuable comments and suggestions in a current subgroup. I am especially pleased to have a junior coworker, Br. Yosuke Miyazaki, who took over the iron project.

I am very happy to work with friendly colleagues, Dr. Hiroki Hanayama, Dr. Toshiki Shimizu, Dr. Ryosuke Sekine, Mr. Ko Kamei, and Mr. Takumi Sakamaki. They always motivated me and gave me enjoyable daily life in Nakamura Lab. I would also like to show my appreciation to all other members in Nakamura Lab.

I thank the Program of Leading Graduate Schools (MERIT) for financial support and giving various opportunities to learn from both academia and industry. All of my MERIT colleagues are appreciated, who are always inspiring me from other research fields of chemistry and physics. I also thank the Japan Society for Promotion of Science (JSPS) Research Fellowship for Young Scientists for financial support.

Finally, I would like to express my deep appreciation to my parents for their constant assistance and affectionate encouragement.

February 2022 Takahiro Doba

List of Publications

1. "Nakamura, E. Homocoupling-Free Iron-Catalysed Twofold C–H Activation/ Cross-Couplings of Aromatics via Transient Connection of Reactants"
   Takahiro Doba, Tatsuaki Matsubara, Laurean Ilies, Rui Shang, Eiichi Nakamura
   Nat. Catal. **2019**, *2*, 400–406.
   (Highlighted in Nature Portfolio Chemistry Community, EurekAlert, Alpha Galileo, and Chem-station, Press release)
2. "Chromium(III)-Catalyzed C(sp^2)–H Alkynylation, Allylation, and Naphthalenation of Secondary Amides with Trimethylaluminum as Base"
   Mengqing Chen, Takahiro Doba, Takenari Sato, Hlib Razumkov, Laurean Ilies, Rui Shang, Eiichi Nakamura
   J. Am. Chem. Soc. **2020**, *142*, 4883–4891.
3. "Iron-Catalysed Regioselective Thienyl C–H/C–H Coupling"
   Takahiro Doba, Laurean Ilies, Wataru Sato, Rui Shang, Eiichi Nakamura
   Nat. Catal. **2021**, *4*, 631–638.
   (Highlighted in Nature Portfolio Chemistry Community)
4. "Triarylamine/Bithiophene Copolymer with Enhanced Quinoidal Character as Hole-Transporting Material for Perovskite Solar Cells"
   Hao-Sheng Lin‡, Takahiro Doba‡, Wataru Sato, Yutaka Matsuo, Rui Shang, Eiichi Nakamura
   Angew. Chem. Int. Ed. **2022**, *61*, e202203949.
   (‡ Equal contribution, Selected as a Hot Paper)
5. "Iron-Catalyzed C–H Activation for Heterocoupling and Copolymerization of Thiophenes with Enamines"
   Takahiro Doba, Rui Shang, Eiichi Nakamura
   J. Am. Chem. Soc. **2022**, *144*, 21692–21701.
6. "Versatile Synthesis of Trisphosphines Bearing Phenylene and Vinylene Backbones Useful for Metal Catalysis and Materials Applications"
   Takahiro Doba, Shota Fukuma, Rui Shang, Eiichi Nakamura
   Synthesis **2023**, *54*, 1690–1699.

Contents

Chapter 1
General Introduction

1.1 Transition-Metal-Catalyzed C–H/C–H Coupling

Conjugated compounds containing C(sp^2)–C(sp^2) linkages are one of the most important classes of compounds because of the interaction of their *p*-orbitals with electrons and photons [1]. Therefore, development of C(sp^2)–C(sp^2) bond-forming reactions is an important subject in organic chemistry. One of the most widely used methods is transition-metal-catalyzed cross-coupling which has been investigated extensively since 1970s [2] and for which the Nobel Prize in Chemistry 2010 [3] was awarded. Although this methodology is applicable to a broad range of substrates, it has an inevitable limitation that prefunctionalization of both substrates is required, rendering multiple steps for the synthesis of starting materials and special care in handling unstable substrates, especially heteroaryl organometallic reagents and halides [4] (Scheme 1.1).

To overcome these problems, transition-metal-catalyzed C–H/C–H coupling which merges two C–H bonds to form a C–C bond, has attracted considerable attention as an alternative method [5]. However, there are many hurdles to be overcome to realize this ideal transformation. First, C–H bond is more inert than C–M and C–X bonds in terms of bond dissociation energy [6]. In addition, distinguishing multiple C–H bonds present in the molecules is challenging, which directly affects the regioselectivity and chemoselectivity (Scheme 1.2).

Currently, these issues are solved by choosing an appropriate combination of substrates and the modes of the C–H activation processes. Regarding the choice of substrates, the utilization of electronic bias or steric effect has been a classical solution. In 1993, a groundbreaking discovery was done by Murai and coworkers [7] in which they introduced a paradigm called "directing group strategy." A coordinating group enhances the reaction efficiency and selectivity by assisting catalyst to approach the proximal C–H bond to form a metallacycle intermediate (Scheme 1.3).

T. Doba, *Iron-Catalyzed C–H/C–H Coupling for Synthesis of Functional Small Molecules and Polymers*, Springer Theses,
https://doi.org/10.1007/978-981-99-4121-6_1

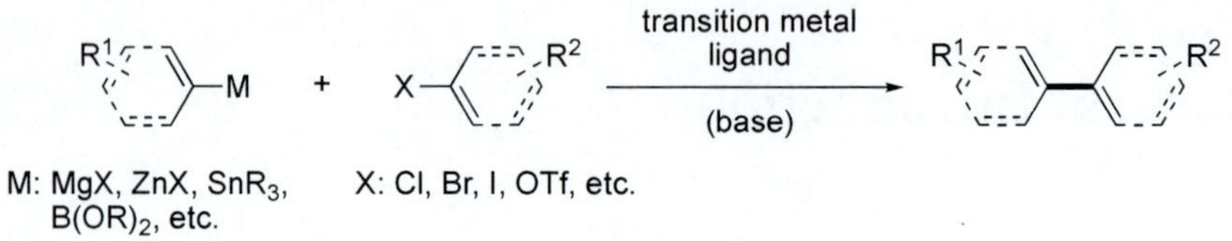

Scheme 1.1 Transition-metal-catalyzed cross-coupling

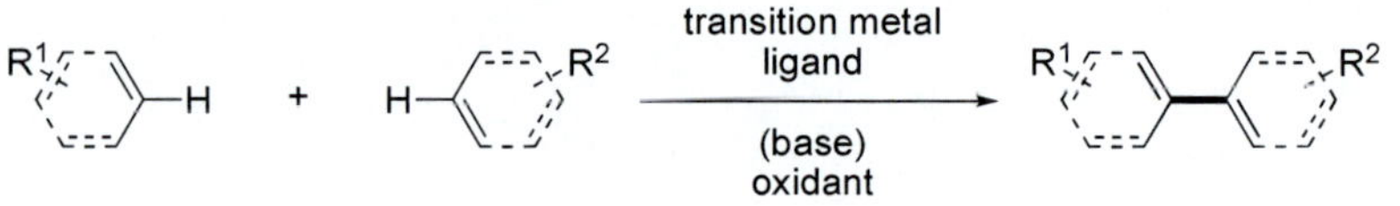

Scheme 1.2 Transition-metal-catalyzed C–H/C–H coupling

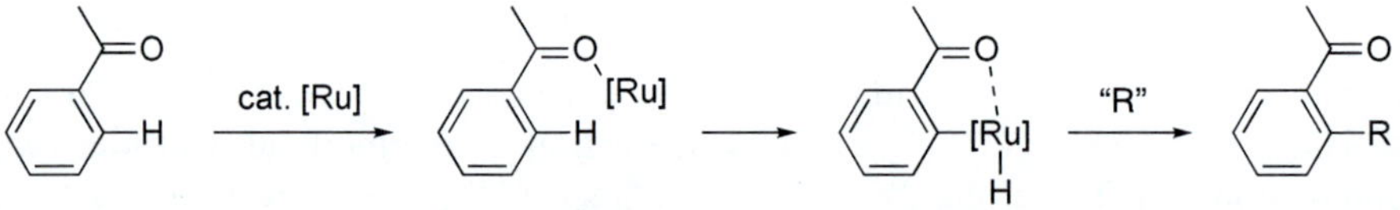

Scheme 1.3 Directing group strategy introduced by Murai and coworkers with ruthenium

Regarding the choice of the modes of the C–H activation processes, electrophilic aromatic substitution (S_EAr) [8], concerted-metalation-deprotonation (CMD) [9], σ-bond metathesis [10], and oxidative addition [11] have been proposed as general modes. Control of the C–H activation modes by careful selection of a catalytic system allows us to achieve efficient and selective transition-metal-catalyzed C–H/C–H coupling. For example, an S_EAr pathway takes place at the most nucleophilic position, while CMD and σ-bond metathesis pathways take place mainly at the most acidic C–H bond. An oxidative addition pathway is often incompatible with the C–H/C–H coupling catalytic cycle because the oxidation state of catalyst changes before and after the C–H activation event (Fig. 1.1).

Currently, transition-metal-catalyzed C–H/C–H coupling reactions heavily rely on the reactivity of palladium and rhodium. Since Fagnou and coworkers reported a pioneering work using palladium as a catalyst, many researchers have investigated this reaction system to broaden the scope [12]. It has been reported that rhodium catalysis is especially effective for the reaction of heteroarenes [13] (Scheme 1.4).

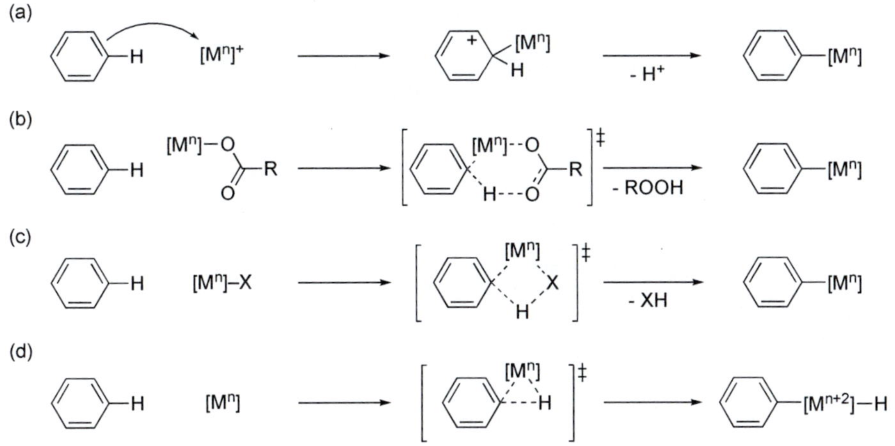

Fig. 1.1 General modes for transition-metal-catalyzed C–H activation processes. **a** electrophilic aromatic substitution (S_EAr), **b** concerted-metalation-deprotonation (CMD), **c** σ-bond metathesis, **d** oxidative addition

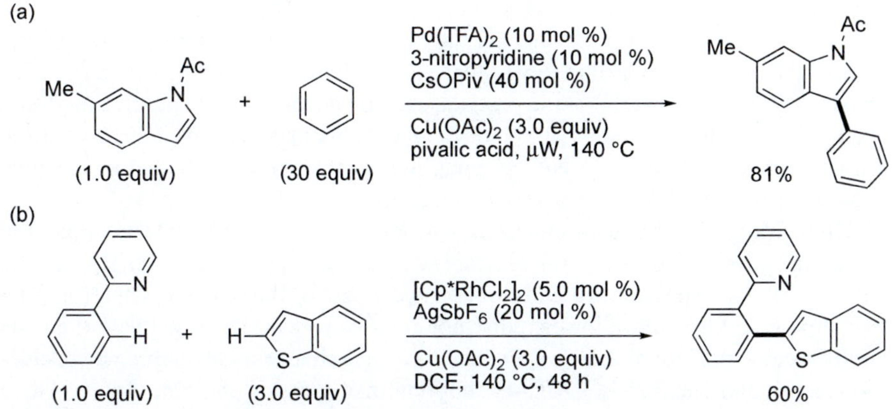

Scheme 1.4 Examples of transition-metal-catalyzed C–H/C–H coupling

1.2 Requirement of Strong Oxidant in Transition-Metal-Catalyzed C–H/C–H Coupling

Because transition-metal-catalyzed C–H/C–H coupling is a coupling of two nucleophiles, transition-metal-catalyzed C–H/C–H coupling requires the addition of an external oxidant for the catalyst turnover [5]. Figure 1.2 shows the catalytic cycle of transition-metal-catalyzed C–H/C–H coupling. First, transition-metal catalyst **I**

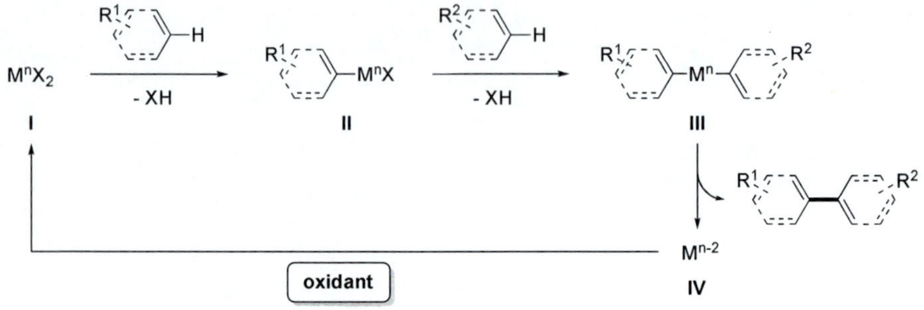

Fig. 1.2 Requirement of oxidant in transition-metal-catalyzed C–H/C–H coupling

activates the C(sp^2)–H bond twice without changing its oxidation state to give intermediate **III** bearing two coupling fragments. Then **III** undergoes reductive elimination to afford the C–H/C–H coupling product and a transition-metal catalyst whose oxidation state is reduced by two (**IV**). To oxidize the low-valent metal species **IV** to high-valent metal species **I**, addition of a stoichiometric oxidant is necessary.

Currently, transition-metal-catalyzed C–H/C–H coupling requires the use of strong oxidants such as Ag(I), Cu(II), *p*-benzoquinone (BQ), and oxygen.[5] The use of strong oxidant in transition-metal-catalyzed C–H/C–H coupling is problematic because the oxidant may oxidize not only the low-valent transition-metal catalyst (**IV**) but also the starting material or the product, causing decomposition or unwanted side reactions [14]. This situation makes it difficult to apply this attractive transformation, transition-metal-catalyzed C–H/C–H coupling, to the synthesis of redox-sensitive π-materials of importance in materials science.

The rationale for the requirement of a strong oxidant is the high redox potential of the transition-metal catalyst. For example, the redox potential of Pd(II)/Pd(0) is 0.915 V versus NHE [15], which limits the weakness of the oxidant. Therefore, if a transition metal that has a lower redox potential can successfully activate the C–H bond twice (from **I** to **IV** in Fig. 1.2), there is a high chance of developing a versatile transition-metal-catalyzed C–H/C–H coupling that operates under mildly oxidative conditions and enables the synthesis of redox-sensitive π-materials of importance in materials science. Aiming for this ultimate goal, I envisioned to utilize the reactivity of iron.

1.3 Iron-Catalyzed Directed C–H Activation

Iron is the final product of stellar nucleosynthesis and therefore is the most abundant transition metal on earth [16]. Nakamura and coworkers have reported iron-catalyzed C–H activation taking advantage of its high abundance, low cost, and low toxicity [17]. However, iron can take various oxidation states and spin states, making it

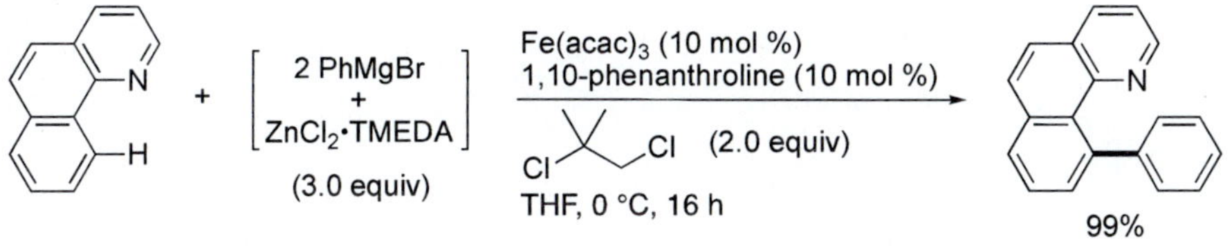

Scheme 1.5 The first example of iron-catalyzed C–H activation reaction reported by Yoshikai and coworkers in 2008

difficult to express a desired reactivity [18]. In 2008, Yoshikai and coworkers reported that a bipyridine ligand in combination with a pyridyl directing group can effectively control the reactivity of iron to catalyze C–H arylation with arylzinc reagents [19]. Arylzinc reagents served not only as a coupling partner but also as a base for C–H activation through σ-bond metathesis and dihaloalkane was determined as an optimal oxidant. The reaction was conducted at low temperature (as low as 0 °C) to prevent the degradation of unstable organoiron intermediates (Scheme 1.5).

In 2013, Nakamura and coworkers reported that a combination of a bidentate phosphine ligand and a bidentate directing group [20] is beneficial for efficient C–H activation and stabilization of the resulting iron metallacycle intermediate, which can react with various kinds of organometallic nucleophiles [21] and electrophiles such as allyl ether [22], *N*-chloroamine [23], alkyl (pseudo)halide [24], and olefins [25]. The existence of an iron metallacycle intermediate was confirmed by deuterium quenching experiments and more recently by direct isolation of the intermediate [26] (Scheme 1.6).

In 2016, Nakamura and coworkers further reported that a combination of a tridentate phosphine ligand and a monodentate carbonyl directing group is also suitable for iron-catalyzed C–H activation [27]. Methylation of weakly coordinating aromatic

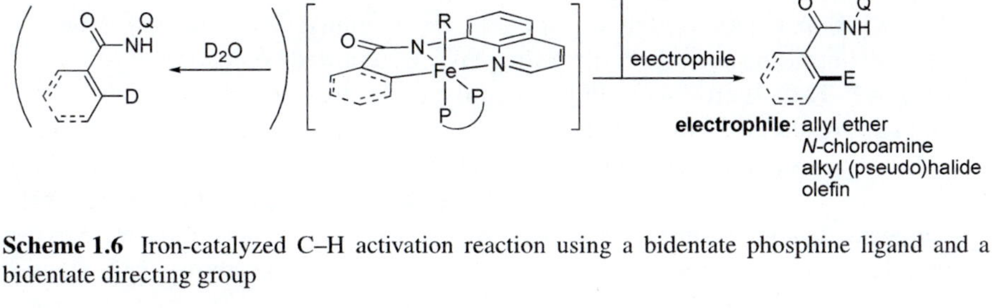

Scheme 1.6 Iron-catalyzed C–H activation reaction using a bidentate phosphine ligand and a bidentate directing group

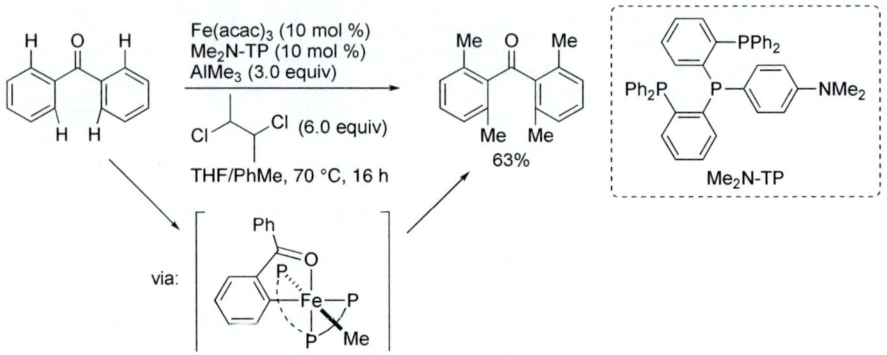

Scheme 1.7 Iron-catalyzed C–H activation reaction using a tridentate phosphine ligand and a monodentate carbonyl directing group

ketones proceeded smoothly using $AlMe_3$ as a methyl source and dihaloalkane as a mild oxidant (Scheme 1.7).

Putting together all of the iron-catalyzed C–H activation reactions described above, dihaloalkanes can be used as a mild oxidant in all cases. This is because the redox potential of Fe(III)/Fe(I) is low and Fe(I) can be easily oxidized to Fe(III) by a mild oxidant to regenerate the catalyst. The redox potential of Fe(III)/Fe(I) is estimated to be lower than 0.55 V *vs* NHE which is lower than that of a conventional Pd(II)/Pd(0) (0.915 V *vs* NHE) cycle [15]. Therefore, an Fe(III)/Fe(I) catalytic cycle with a low redox potential is suitable for achieving C–H/C–H coupling that operates with a mild oxidant and enables the synthesis of redox-sensitive π-materials.

1.4 Iron-Catalyzed Regioselective Thienyl C–H Activation

As seen in the reactions described above, iron-catalyzed C–H activation reactions were limited to the synthesis of *ortho*-disubstituted arenes possessing a coordinating group for the sake of a directing group strategy. The obtained products are not always appreciated as a useful compound for practical applications and in many cases removal of a preinstalled directing group is required [28].

In my master course studies, I obtained a foothold for achieving iron-catalyzed C–H activation reactions that do not rely on a preinstalled directing group. In 2019, Nakamura and coworkers (including myself) reported homocoupling-free iron-catalyzed C–H/C–H cross-coupling of carboxamides with thiophenes (Scheme 1.8) [29]. In this reaction, I found that the most acidic C–H bond of thiophene compounds that do not possess an extraneous directing group can be activated regioselectively by a stable organoiron species (as shown in Scheme 1.6) through a σ-bond metathesis mechanism. The mechanisms of iron-catalyzed thienyl C–H activation were experimentally supported by deuterium exchange between two substrates. This reaction

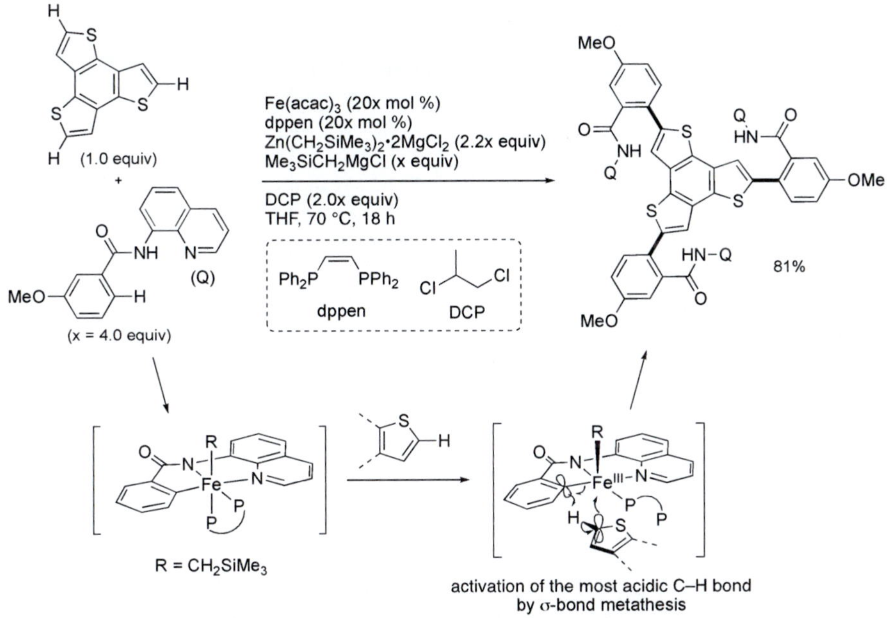

Scheme 1.8 Homocoupling-free iron-catalyzed C–H/C–H cross-coupling of carboxamides with thiophenes

mechanism was the key to completely suppress the formation of homocoupling products. Although a preinstalled bidentate directing group was necessary for one of the substrates, the reaction was applicable to various kinds of thiophene cores commonly found in real optoelectronic materials [30].

1.5 Objective and Outline of This Thesis

The application of transition-metal-catalyzed C–H/C–H coupling, an ideal coupling method for the construction of a C(sp^2)–C(sp^2) bond, to the synthesis of redox-sensitive π-materials was hindered by the requirement of a strong oxidant to turn over the catalyst. To overcome this problem, my Ph.D. studies focused on the low redox potential of Fe(III)/Fe(I), enabling C–H/C–H coupling under mildly oxidative conditions. Thiophene compounds were chosen as substrates because of the wide application found in materials science. As revealed by my master course studies, the electronic bias created by a sulfur atom helps to make the adjacent C–H bond acidic enough for iron-catalyzed C–H activation through a σ-bond metathesis mechanism. The reactions described in this thesis do not require a preinstalled directing group, enabling direct application of the obtained products as organic materials.

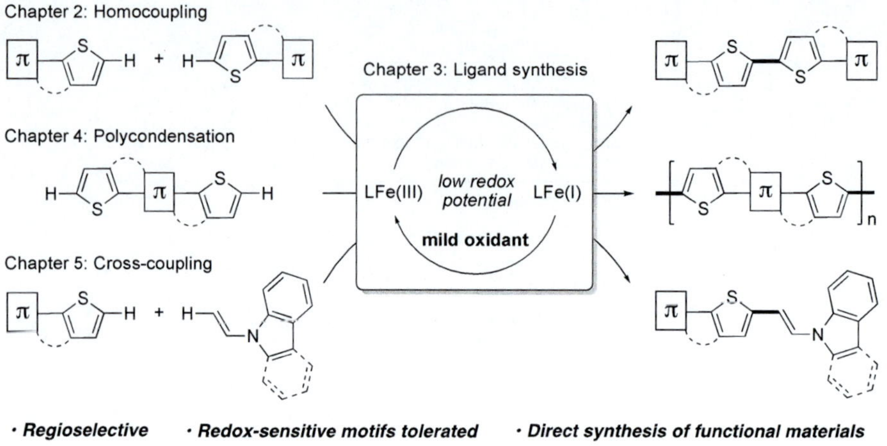

Fig. 1.3 Iron-catalyzed regioselective thienyl C–H/C–H coupling for synthesis of functional small molecules and polymers

To this end, I first investigated the iron catalytic system that catalyzes regioselective C–H/C–H homocoupling of thiophene compounds and determined that oxalate serves as a mild oxidant in combination with $AlMe_3$ base and a tridentate phosphine ligand. Then, a modular synthetic method of a tridentate phosphine ligand was developed to further accelerate the exploration of this reaction system. To expand the applicability of iron-catalyzed regioselective thienyl C–H/C–H coupling to the synthesis of polymeric compounds, iron-catalyzed regioselective thienyl C–H/C–H polycondensation was developed by suppressing catalyst deactivation by ligand design. The obtained polymer was applied as a hole-transporting material for perovskite solar cells to demonstrate the practicability of this method. Finally, the Fe(III)/trisphosphine/$AlMe_3$/oxalate system was applied to C–H/C–H cross-coupling of thiophene compounds with vinylcarbazoles and vinylindoles, which gives direct access to donor materials containing a vinylthiophene structure. The reactions described herein highlight the potential of iron, the most abundant transition metal on earth, for the direct synthesis of functional small molecules and polymers of importance in materials science (Fig. 1.3).

References

1. (a) Cornil J, Beljonne D, Calbert J-P, Brédas J-L (2001) Adv Mater 13:1053–1067. (b) Wu W, Liu Y, Zhu D (2010) Chem Soc Rev 39:1489–1502. (c) Wang C, Dong H, Hu W, Liu Y, Zhu D (2012) Chem Rev 112:2208–2267. (d) Gierschner J, Cornil J, Egelhaaf H-J (2007) Adv Mater 19:173–191. (e) Hughes G, Bryce MRJ (2005) Mater Chem 15:94–107. (f) Tour JM (1996) Chem Rev 96:537–554. (g) Adachi C (2014) Jpn J Appl Phys 53:060101. (h) Kanis DR, Ratner MA, Marks TJ (1994) Chem Rev 94:95–242. (i) Cai J, Ruffieux P, Jaafar R, Bieri M, Braun T, Blankenburg S, Muoth M, Seitsonen AP, Saleh M, Feng X, Müllen K, Fasel R (2010) Nature

466:470–473. (j) Ruffieux P, Wang S, Yang B, Sánchez-Sánchez C, Liu J, Dienel T, Talirz L, Shinde P, Pignedoli CA, Passerone D, Dumslaff T, Feng X, Müllen K, Fasel R (2016) Nature 531:489–492

2. (a) Tamura M, Kochi JK (1971) J Am Chem Soc 93:1487–1489. (b) Tamao K, Sumitani K, Kumada M (1972) J Am Chem Soc 94:4374–4376. (c) Corriu RJP, Masse JPJ (1972) Chem Soc, Chem Commun (3):144. (d) Mizoroki T, Mori K, Ozaki A (1971) Bull Chem Soc Jpn 44:581–581. (e) Heck RF, Nolley JP (1972) J Org Chem 37:2320–2322. (f) Yamamura M, Moritani I, Murahashi S-I (1975) J Organometal Chem 91:C39–C42. (g) Sonogashira K, Tohda Y, Hagihara N (1975) Tetrahedron Lett 16:4467–4470. (h) King AO, Okukado N, Negishi E (1977) J Chem Soc, Chem Commun (19):683–684. (i) Kosugi M, Sasazawa K, Shimizu Y, Migita T (1977) Chem Lett 6:301–302. (j) Milstein D, Stille JK (1978) J Am Chem Soc 100:3636–3638. (k) Miyaura N, Suzuki A (1979) J Chem Soc, Chem Commun (19):866–867. (l) Hatanaka Y, Hiyama T (1988) J Org Chem 53:918–920
3. Palladium-Catalyzed Cross-Coupling in Organic Synthesis. https://www.nobelprize.org/uploads/2018/06/advanced-chemistryprize2010-1.pdf
4. Kinzel T, Zhang Y, Buchwald SL (2010) J Am Chem Soc 132:14073–14075
5. Yang Y, Lan J, You J (2017) Chem Rev 117:8787–8863
6. McMillen DF, Golden DM (1982) Annu Rev Phys Chem 33:493–532
7. Murai S, Kakiuchi F, Sekine S, Tanaka Y, Kamatani A, Sonoda M, Chatani N (1993) Nature 366:529–531
8. (a) Pivsa-Art S, Satoh T, Kawamura Y, Miura M, Nomura M (1998) Bull Chem Soc Jpn 71:467–473. (b) Lane BS, Brown MA, Sames D (2005) J Am Chem Soc 127:8050–8057
9. (a) García-Cuadrado D, de Mendoza P, Braga AAC, Maseras F, Echavarren AM (2007) J Am Chem Soc 129:6880–6886. (b) Gorelsky SI, Lapointe D, Fagnou K (2008) J Am Chem Soc 130:10848–10849. (c) Gorelsky SI, Lapointe D, Fagnou K (2012) J Org Chem 77:658–668
10. (a) Thompson ME, Baxter SM, Bulls AR, Burger BJ, Nolan MC, Santarsiero BD, Schaefer WP, Bercaw JE (1987) J Am Chem Soc 109:203–219. (b) Hennessy EJ, Buchwald SL (2003) J Am Chem Soc 125:12084–12085
11. (a) Janowicz AH, Bergman RG (1982) J Am Chem Soc 104:352–354. (b) Chen H, Schlecht S, Semple TC, Hartwig JF (2000) Science 287:1995–1997. (c) Cho J-Y, Tse MK, Holmes D, Maleczka RE, Smith MR (2002) Science 295:305–308. (d) Ishiyama T, Takagi J, Ishida K, Miyaura N, Anastasi NR, Hartwig JF (2002) J Am Chem Soc 124:390–391
12. Stuart DR, Fagnou K (2007) Science 316:1172–1175
13. Dong J, Long Z, Song F, Wu N, Guo Q, Lan J, You J (2013) Angew Chem Int Ed 52:580–584
14. Grzybowski M, Skonieczny K, Butenschön H, Gryko DT (2013) Angew Chem Int Ed 52:9900–9930
15. Shriver D, Atkins P, Overton T, Rourke J, Weller M, Armstrong F (2006) Shriver & Atkins inorganic chemistry. Oxford University Press, Oxford
16. (a) Hoyle F (1946) Monthly Notices Roy Astron Soc 106:343–383. (b) Burbidge EM, Burbidge GR, Fowler WA, Hoyle F (1957) Rev Mod Phys 29:547–650.
17. (a) Nakamura E, Sato K (2011) Nature Mater 10:158–161. (b) Shang R, Ilies L, Nakamura E (2017) Chem Rev 117:9086–9139.
18. Sun Y, Tang H, Chen K, Hu L, Yao J, Shaik S, Chen H (2016) J Am Chem Soc 138:3715–3730
19. Norinder J, Matsumoto A, Yoshikai N, Nakamura E (2008) J Am Chem Soc 130:5858–5859
20. Zaitsev VG, Shabashov D, Daugulis O (2005) J Am Chem Soc 127:13154–13155
21. (a) Shang R, Ilies L, Matsumoto A, Nakamura EJ (2013) Am Chem Soc 135:6030–6032. (b) Shang R, Ilies L, Asako S, Nakamura EJ (2014) Am Chem Soc 136:14349–14352. (c) Shang R, Ilies L, Nakamura EJ (2015) Am Chem Soc 137:7660–7663. (d) Ilies L, Ichikawa S, Asako S, Matsubara T, Nakamura E (2015) Adv Synth Catal 357:2175–2179. (e) Ilies L, Itabashi Y, Shang R, Nakamura E (2017) ACS Catal 7:89–92.
22. Asako S, Ilies L, Nakamura E (2013) J Am Chem Soc 135:17755–17757
23. Matsubara T, Asako S, Ilies L, Nakamura E (2014) J Am Chem Soc 136:646–649
24. Ilies L, Matsubara T, Ichikawa S, Asako S, Nakamura E (2014) J Am Chem Soc 136:13126–13129

25. (a) Matsubara T, Ilies L, Nakamura E (2016) Chem. Asian J 11:380–384. (b) Ilies L, Zhou Y, Yang H, Matsubara T, Shang R, Nakamura E (2018) ACS Catal 8:11478–11482. (c) Ilies L, Arslanoglu Y, Matsubara T, Nakamura E (2018) Asian J Org Chem 7:1327–1329
26. Boddie TE, Carpenter SH, Baker TM, DeMuth JC, Cera G, Brennessel WW, Ackermann L, Neidig ML (2019) J Am Chem Soc 141:12338–12345
27. Shang R, Ilies L, Nakamura E (2016) J Am Chem Soc 138:10132–10135
28. Kuhl N, Hopkinson MN, Wencel-Delord J, Glorius F (2012) Angew Chem Int Ed 51:10236–10254
29. Doba T, Matsubara T, Ilies L, Shang R, Nakamura E (2019) Nat Catal 2:400–406
30. (a) Shirota Y, Kageyama H (2007) Chem Rev 107:953–1010. (b) Wang C, Dong H, Hu W, Liu Y, Zhu D (2012) Chem Rev 112:2208–2267. (c) Sirringhaus H (2014) Adv Mater 26:1319–1335. (d) Facchetti A (2007) Mater Today 10:28–37. (e) Wu W, Liu Y, Zhu D (2010) Chem Soc Rev 39:1489–1502. (f) O'Neill M, Kelly SM (2011) Adv Mater 23:566–584. (g) Murphy AR, Fréchet JMJ (2007) Chem Rev 107:1066–1096. (h) Zhang F, Wu D, Xu Y, Feng X (2011) J Mater Chem 21:17590. (i) Ameri T, Khoram P, Min J, Brabec (2013) Adv Mater 25:4245–4266

Chapter 2
Iron-Catalyzed Regioselective Thienyl C–H/C–H Homocoupling

2.1 Introduction

Bithiophene compounds have wide application in the field of organic electronics such as organic field-effect transistors (OFETs), organic light-emitting diodes (OLEDs), and organic solar cells (OSCs) [1]. Aiming for the direct synthesis of bithiophene compounds, transition-metal-catalyzed thienyl C–H/C–H homocoupling has attracted much attention as one of the most straightforward methods to synthesize those compounds from simple thiophene substrates. The first example of transition-metal-catalyzed thienyl C–H/C–H homocoupling was reported by Kozhevnikov in 1976 using $Pd(OAc)_2$ as a catalyst and $Cu(OAc)_2$ as an oxidant (Scheme 2.1) [2]. This reaction gave a mixture of 2,2′-bithiophene and 2,3′-bithiophene.

Despite the synthetic significance of transition-metal-catalyzed thienyl C–H/C–H homocoupling, it is only recently that a synthetically useful versions have been investigated seriously. In 2004, Mori and coworkers succeeded in reducing the amount of thiophene starting materials to 1 equiv in palladium-catalyzed thienyl C–H/C–H homocoupling (Scheme 2.2) [3]. The key was the use of AgF both as an oxidant and an effective promoter to generate an active catalytic intermediate.

After this report, several other palladium catalytic systems were reported. In 2019, Carrow and coworkers reported that a thioether ligand facilitates the palladium-catalyzed thienyl C–H/C–H homocoupling by *electrophilic* concerted metalation-deprotonation (*e*CMD) mechanism where positive charge buildup is observed on the substrate [4] (Scheme 2.3).

Stahl and coworkers also reported the palladium-catalyzed aerobic thienyl C–H/C–H homocoupling using phenanthroline dione (phd) as an ancillary ligand and $Cu(OAc)_2$ as a cocatalyst [5]. They ascribed the superior performance of the phd ligand to the formation of a phd-bridged bimetallic species, $(AcO)_2Cu(phd)Pd(OAc)_2$ (Scheme 2.4).

As explained above, a modern palladium-catalyzed C–H activation has broadened the scope, realizing a simple synthetic route to dimeric thiophene compounds.

T. Doba, *Iron-Catalyzed C–H/C–H Coupling for Synthesis of Functional Small Molecules and Polymers*, Springer Theses,
https://doi.org/10.1007/978-981-99-4121-6_2

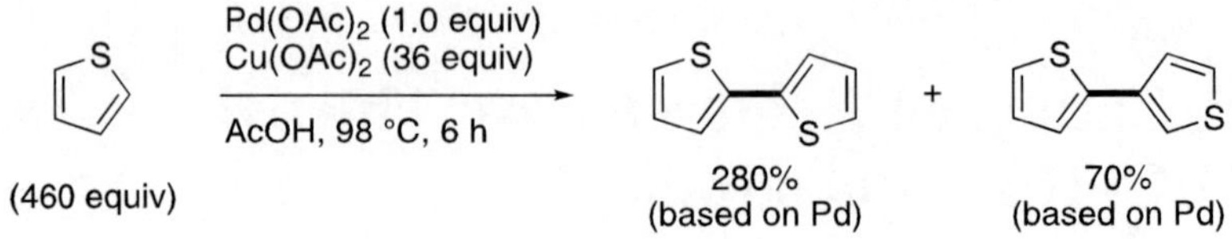

Scheme 2.1 First example of transition-metal-catalyzed thienyl C–H/C–H homocoupling

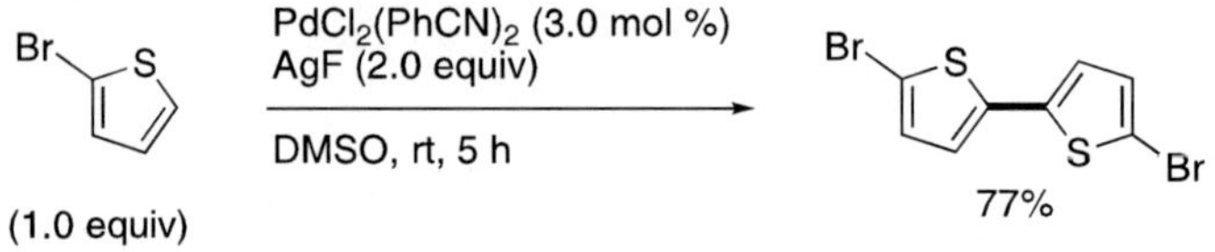

Scheme 2.2 Palladium-catalyzed thienyl C–H/C–H homocoupling reported by Mori and coworkers

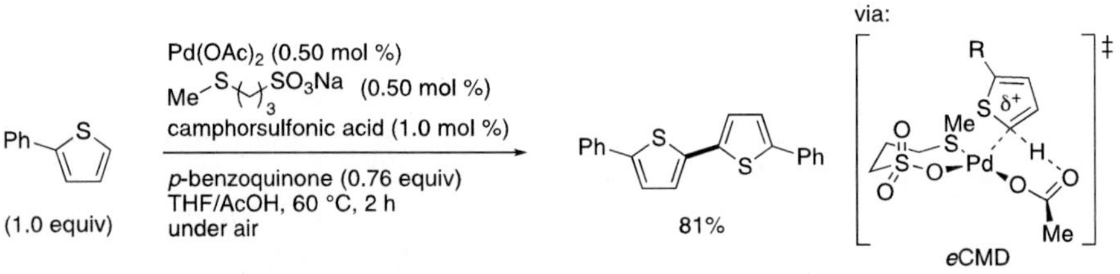

Scheme 2.3 Palladium-catalyzed thienyl C–H/C–H homocoupling through eCMD process

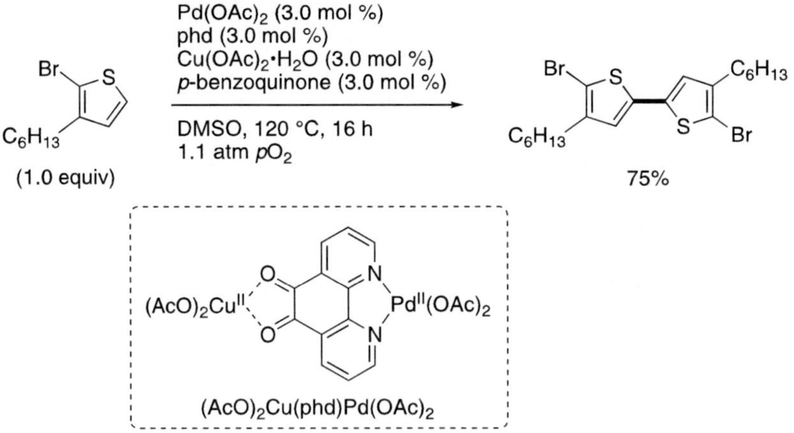

Scheme 2.4 Palladium-catalyzed aerobic thienyl C–H/C–H homocoupling

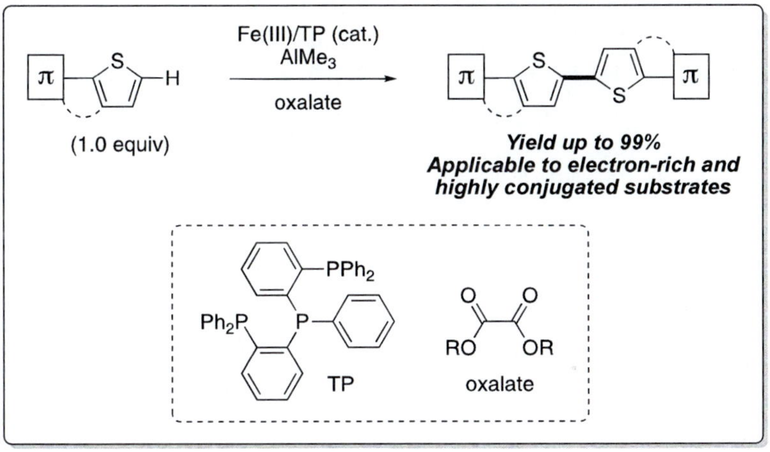

Scheme 2.5 Iron-catalyzed regioselective thienyl C–H/C–H homocoupling

However, these reactions require strongly oxidizing conditions to turn over the Pd(II)/Pd(0) cycle with a large redox potential [$E°$ (Pd^{II}/Pd^{0}) = + 0.915 V *vs* NHE], which limits substrate versatility and reaction selectivity, especially in the synthesis of electron-rich and highly conjugated molecules. In this chapter, I report iron-catalyzed regioselective thienyl C–H/C–H homocoupling using conjugated tridentate phosphine as a ligand, $AlMe_3$ as a base, and diethyl oxalate as a mild oxidant, enabling the synthesis of π-conjugated dimeric and oligomeric thiophene compounds of importance in materials science (Scheme 2.5).

2.2 Initial Discovery

The starting point of the investigation was the iron-catalyzed *ortho* C–H methylation of aromatics bearing a simple carbonyl group as a directing group reported by Nakamura and coworkers in 2016 (Scheme 2.6) [6]. The key to the success was to use tridentate phosphine as a ligand and $AlMe_3$ as a mild base and a methyl donor to generate organoiron species stable enough to activate the C–H bond of weakly coordinating substrates. As seen in the crystal structure of Fe(II)/TP complexes [7], three phosphine groups are suitable for tridentate coordination. $AlMe_3$ was solely effective, and other organometallic reagents such as $AlEt_3$, $AlPh_3$, $ZnMe_2$, and MeMgBr were all ineffective.

Based on these results, I expected that this reaction system is also applicable to the C–H activation of thiophene compounds that do not possess a directing group, which is also weakly coordinating through the sulfur atom or the π orbitals [8]. To my delight, using benzo[*b*]thiophene as a substrate, the desired homocoupling product was obtained in moderate yield by regioselective C–H activations at the most acidic

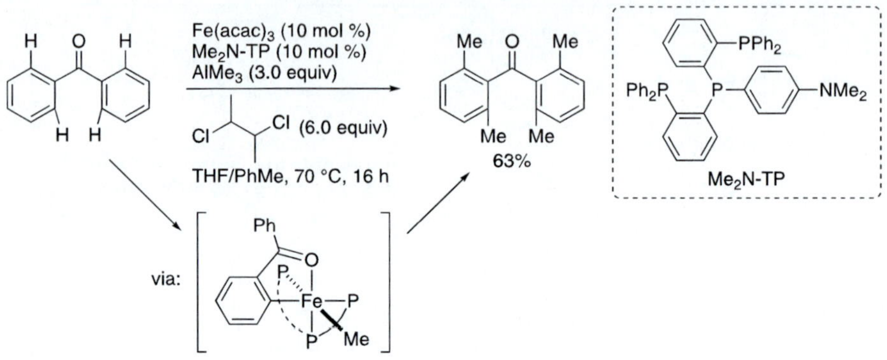

Scheme 2.6 Iron-catalyzed ortho C–H methylation of aromatics bearing a simple carbonyl group as a directing group

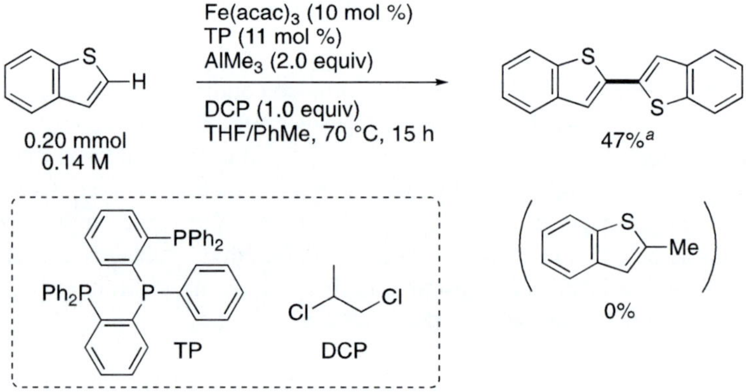

[a] The yield was determined by ^{1}H NMR using 1,1,2,2-tetrachloroethane as an internal standard.

Scheme 2.7 Initial discovery of iron-catalyzed regioselective thienyl C–H/C–H homocoupling

C–H bond next to the sulfur atom (Scheme 2.7) [9]. Notably, no methylation product was detected probably because of the slow reductive elimination with the methyl group [10]. However, even after rigorous investigation of the reaction parameters, the yield reached no higher than 50%.

2.3 Catalyst Poisoning by Alkene

Taking into account that the yield of the homocoupling product always stopped at around 50%, I speculated that propene generated by reduction of DCP is inhibiting the reaction by catalyst poisoning (Scheme 2.8).

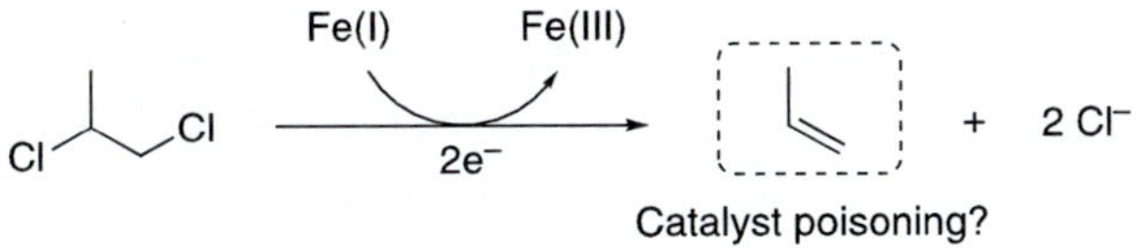

Scheme 2.8 Generation of propene by reduction of DCP

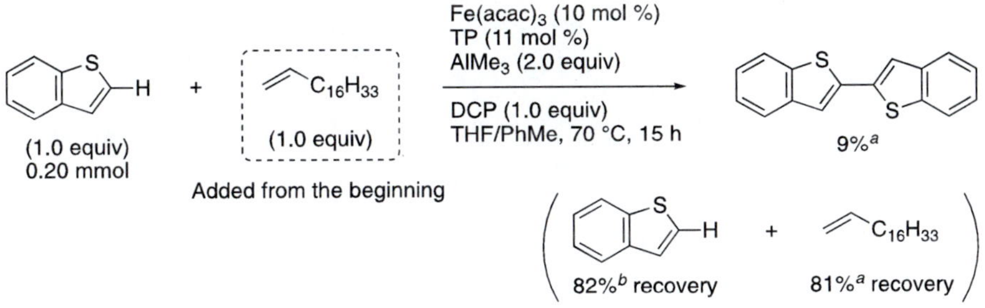

[a] Yields were determined by ^{1}H NMR using 1,1,2,2-tetrachloroethane as an internal standard.
[b] The yield was determined by GC using tridecane as an internal standard.

Scheme 2.9 Inhibition effect of alkene

To investigate the inhibition effect of alkene, 1-octadecene was added from the beginning of the reaction (Scheme 2.9). As expected, the catalyst failed to turnover, giving only 9% of the homocoupling product with recovery of benzo[*b*]thiophene and 1-octadecene in 82% and 81%, respectively. This result clearly shows that the reaction is inhibited by the coordination of alkene to the catalyst when DCP is used as an oxidant. This is in stark contrast with iron-catalyzed directed C–H activation [11], where substrates have a stronger coordinating ability than alkenes so that the efficiency of the reactions are not affected by a growing amount of alkene. Therefore, in order to further improve the efficiency of the homocoupling reaction, investigation of other oxidants that do not generate alkenes after its reduction is necessary.

2.4 Diketone as a Mild Oxidant

By knowing that a carbonyl group can coordinate to iron and accept electron in the iron-catalyzed *ortho* C–H methylation reaction, I came up with the idea of using diketone as a mild oxidant in the iron-catalyzed thienyl C–H/C–H homocoupling reaction. The working hypothesis is shown in Fig. 2.1. First, Fe(III)-Me_2 species (**I**) activates the most acidic C–H bond of a thiophene twice through σ-bond metathesis mechanism to afford the Fe(III)-Ar_2 species (**II**). After reductive elimination, the homocoupling product and Fe(I) species is generated. This low-valent species is oxidized by diketone in combination with $AlMe_3$ via inner-sphere electron transfer

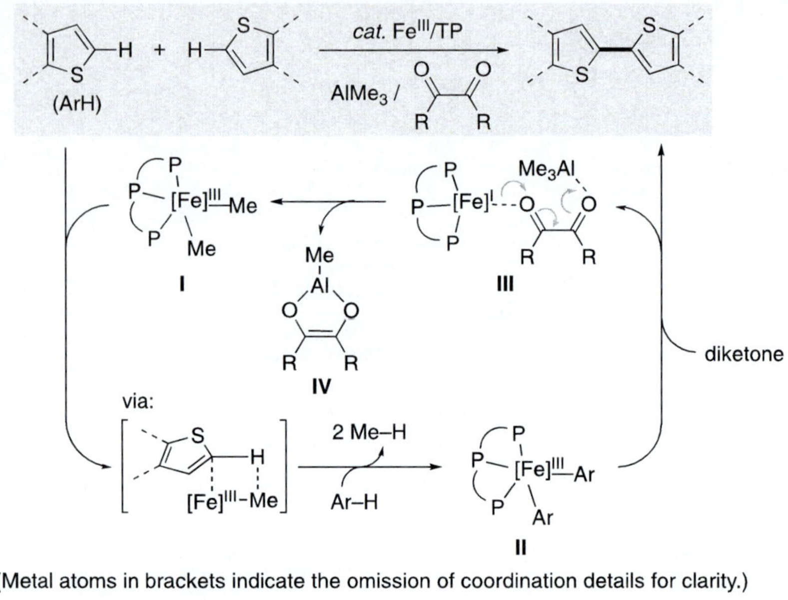

Fig. 2.1 Idea of using diketone as an oxidant

(**III**) to form an aluminum enediolate (**IV**) to close the catalytic cycle. The strong oxophilicity of Al(III) contributes as the driving force for effective catalyst turnover.

Based on this concept, various kinds of diketones were tested as an oxidant (Table 2.1). Among diacetyl, dibenzoyl (benzil), diamide, and diester (oxalate), diester gave the highest conversion of the starting material. No methylation of benzo[*b*]thiophene was observed in any cases. After determining oxalate as a suitable oxidant, the effect of substituents on oxalates was examined. Generally speaking, oxalates with lower LUMO energy gave higher conversion because of higher ability to accept electron, but dimethyl oxalate was inferior to diethyl oxalate. This phenomena can be ascribed to the instability of dimethyl oxalate under the reaction conditions (Scheme 2.10). Diphenyl oxalate also gave lower conversion possibly due to the same reason. By increasing the concentration to 0.7 M, 1.0 equiv of $AlMe_3$ and 0.50 equiv of diethyl oxalate were enough to achieve a full conversion of the starting material. *p*-Benzoquinone often used in palladium-catalyzed C–H activation [12] was ineffective for this reaction. It is a great advantage of using diethyl oxalate as an oxidant because of its mildness, natural abundance, and low cost compared to metal oxidants [e.g., Ag(I), Cu(II)].

Attempts to detect ethyl glyoxalate after hydrolysis or enediolate **IV** in situ have failed so far probably because of the aggregation and high oxophilicity of Al(III) species. However, the reaction using benzil in place of diethyl oxalate produced the homocoupling product in moderate yield and benzoin, providing evidence for the formation of **IV**, and the role of diethyl oxalate as a two-electron acceptor from the

Table 2.1 Investigation of diketone as a mild oxidant

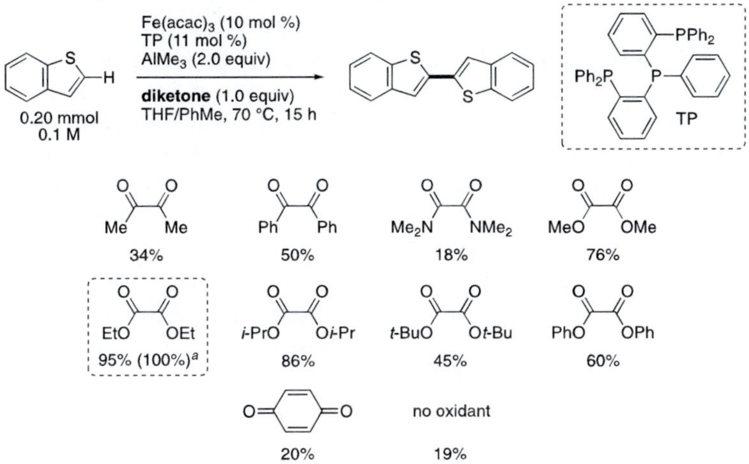

Conversion yields were determined by GC using tridecane as an internal standard.
[a] The reaction was performed with 1.0 equiv of $AlMe_3$ and 0.50 equiv of $(COOEt)_2$ in 0.7 M.

Scheme 2.10 Instability of dimethyl oxalate

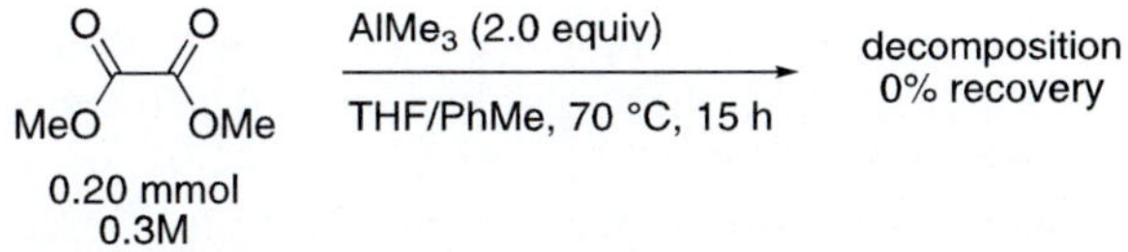

iron catalytic cycle. Moreover, two-electron reduction of diethyl oxalate with alkali metal has been reported [13] (Scheme 2.11).

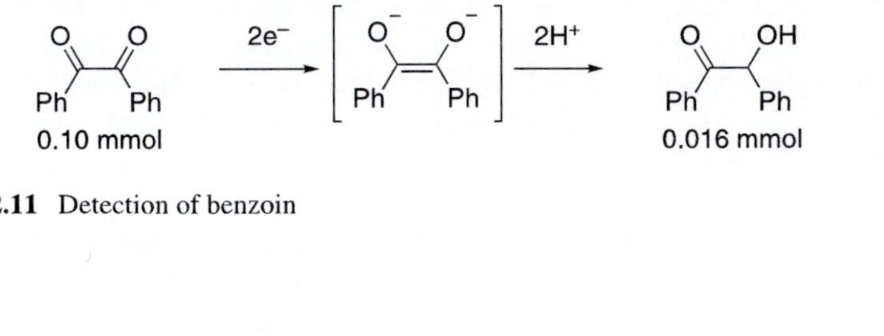

Scheme 2.11 Detection of benzoin

2.5 Effect of Reaction Parameters

With optimal oxidant in hand, I investigated the effect of other reaction parameters on the product yield. Notably, in all cases, no methylation of the starting material was observed and the recovery of the starting material accounted for the mass balance. Under the standard conditions using 10 mol % of $Fe(acac)_3$ as a catalyst, 11 mol % of TP as a ligand, 1.0 equiv of $AlMe_3$ as a base, and 0.50 equiv of diethyl oxalate as an oxidant in a mixed solvent of THF and toluene at 70 °C for 15 h, the reaction proceeded in quantitative yield (Table 2.2, entry 1). A commercially available toluene solution of $AlMe_3$ (2 M) instead of a neat $AlMe_3$ was used to safely operate the organometallic reagent. Control experiments proved that all of the components including THF are necessary (entry 2–6). Ethereal solvent may help to stabilized organoiron species by coordination to the vacant site. The use of Fe(II) instead of Fe(III) significantly decreased the yield, suggesting that the reaction proceeds through a Fe(III)/Fe(I) cycle rather than a Fe(II)/Fe(0) cycle (entry 7). Interestingly, the use of monodentate or bidentate phosphine ligands failed to give the product and iron black was observed, indicating that the stabilization of iron from over-reduction by organometallic base is not sufficient (entry 8 and 9). A tetradentate phosphine ligand, TetraP, also failed to give the product possibly because of the steric hinderance of an additional diphenylphosphino group (entry 10). Other common ligands such as bipyridine, phenanthroline, and *N*-heterocyclic carbene ligands were all ineffective (entry 11–13). Ineffectiveness of $AlEt_3$ was ascribed to the formation of an inactive iron hydride species after β-hydride elimination of the ethyl group (entry 14) [6]. MeLi, MeMgBr, and MeZnBr were also ineffective because of their strong nucleophilicity incompatible with oxalate and high reducing ability (entry 15) [14]. Reaction with less catalyst loading and reaction at lower temperature resulted in incomplete conversions (entry 16 and 17).

2.6 Substrate Scope

Table 2.3 illustrates the scope of iron-catalyzed regioselective thienyl C–H/C–H homocoupling. The reaction cleanly produced the desired product, which was easily isolated either by silica gel chromatography or by filtration and reprecipitation. The reaction took place exclusively at the C–H bond next to the sulfur atom on thiophenes. These results clearly demonstrate that iron-catalyzed C–H activation through σ-bond metathesis mechanism, which favors the most acidic C–H bond, is beneficial to control the regioselectivity of the reaction. C–H methylation instead of C–H/C–H homocoupling barely took place (detected for **15**, **16**, and **17** in $< 1\%$ yield). Dimers of benzo[*b*]thiophene (**2**), 2-phenylthiophene (**3**), and 2,3-disubstituted thiophenes (**4**, **5**) were obtained in high to excellent yields. On the other hand, a hexyl substituent next to the potentially active C–H bond inhibits the C–H activation, producing head-to-head coupled tetrathiophene (**6**) as an exclusive

Table 2.2 Effect of reaction parameters

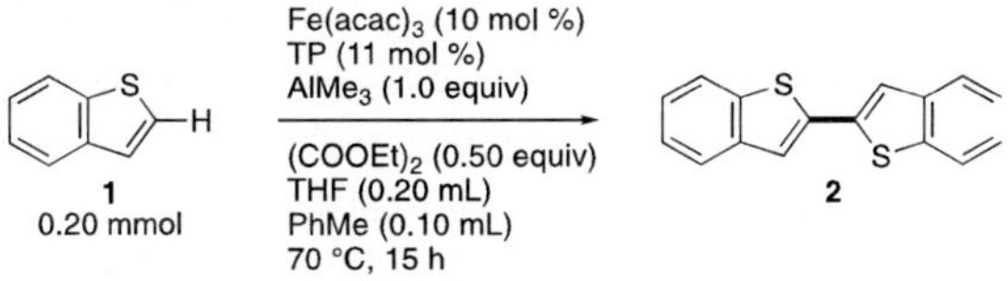

entry	variation from standard condition	yield (%)
1	none	100
2	without $Fe(acac)_3$	0
3	without TP	0
4	without $AlMe_3$	0
5	without $(COOEt)_2$	7
6	without THF	34
7	$Fe(acac)_2$ instead of $Fe(acac)_3$	47
8	PPh_3 (33 mol %) instead of TP	0
9	dppbz instead of TP	0
10	TetraP instead of TP	0
11	bpy instead of TP	0
12	1,10-phen instead of TP	0
13	IMes•HCl instead of TP	0
14	$AlEt_3$ instead of $AlMe_3$	0
15	MeLi, MeMgBr, or MeZnBr instead of $AlMe_3$	0
16	$Fe(acac)_3$ (1.0 mol %), TP (1.1 mol %)	40
17	50 °C instead of 70 °C	35

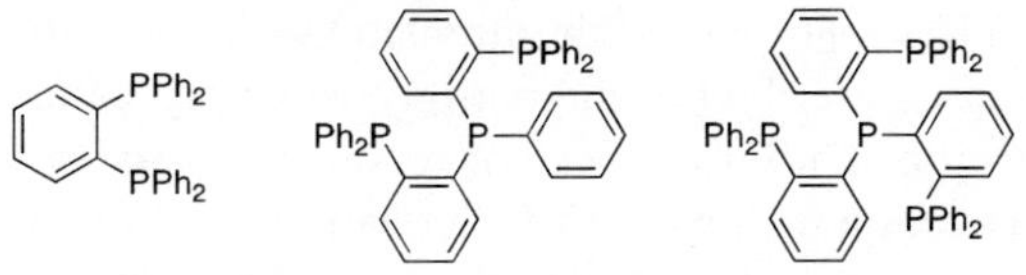

dppbz TP TetraP

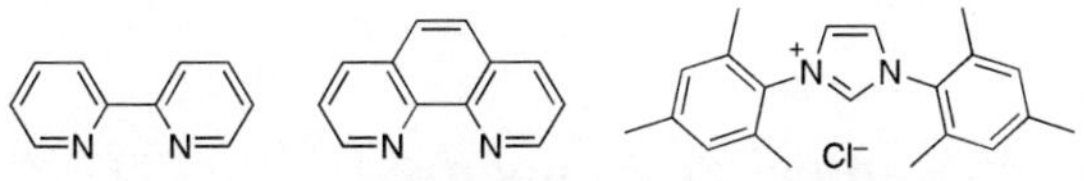

bpy 1,10-phen IMes•HCl

No methylation of **1** was detected by GC.
[a] Yields were determined by ^1H NMR using 1,3,5-trimethoxybenzene as an internal standard.

product. π-Motifs commonly found in optoelectronic materials, such as benzofuran (**7**), benzo[1,2-*b*:4,5-*b*′]dithiophene (**8**), vinylene (**9**), fluorene (**13**), carbazoles (**16**, **17**), and electron-rich triarylamines (**18**, **19**), were well tolerated under mildly oxidative conditions. The reaction tolerates triisopropylsilyl (**8–10**), tributyltin (**11**), and pinacoboronate (**12**) on arenes, which are useful for transition-metal-catalyzed cross-coupling reactions for further functionalization. An oligo-thienylenevinylene product (**9**), which is susceptible to electron-transfer side reactions, was obtained in high yield without isomerization of the double bond. Since 1-octadecene heavily suppressed the reaction (Scheme 2.9), this contrast suggests that the Fe(III)/TP coordination site is sensitive to steric effects of the incoming π-substrates. Electron-rich oligothiophenes with silicon and tin protection at the terminal position were obtained in high yields (**10**, **11**). The non-oxidative conditions created by a mild oxalate oxidant keep intact the multi-aromatic array in **14**, which is potentially susceptible to the conventional oxidative aromatic coupling under strongly oxidizing conditions [15]. Electron-rich 3,4-ethylenedioxythiophene (EDOT) [16] reacted in moderate yield due to steric hindrance by the 3,4-substituents (**15**). Because of the mildness of oxalate as an oxidant for iron, a hole-transporting material for perovskite solar cells (**19**) [17], which is susceptible to direct oxidation by strong oxidant, was synthesized in excellent yield. Reactions under Mori's conditions ($PdCl_2(PhCN)_2$ (3.0 mol %), AgF (2.0 equiv), DMSO, 60 °C, 5 h) only gave the homocoupling products in moderate yields (**2**, **16**), demonstrating the usefulness of the iron homocoupling method.

2.7 Unsuccessful Substrates

Table 2.4 illustrates the examples of unsuccessful substrates. Notably, the efficiency of this iron-catalyzed thienyl C–H/C–H homocoupling significantly drops with thiophene substrates that do not have an extended π conjugation. NICS(1) values [18] calculated at the B3LYP/6-31G(d) level of theory clearly showed that substrates with lower aromaticity are more reactive and substrates with higher aromaticity are less reactive (Fig. 2.2a). As reported in other transition-metal-catalyzed reactions [19], this phenomena can be ascribed to the partial loss of aromaticity in the transition state of C–H activation. With non-π-extended substrates, or substrates with higher aromaticity, the activation energy of C–H activation would be too high because of large dearomatization energy of the thienyl group (Fig. 2.2b).

2.8 Addition of Radical Scavenger

To see if any radical process is taking place, iron-catalyzed regioselective thienyl C–H/C–H homocoupling was conducted in the presence of a radical scavenger, 9,10-dihydroanthracene. As shown in Scheme 2.12, the homocoupling product was obtained in quantitative yield even in the presence of a radical scavenger and the radical scavenger was fully recovered. These results rule out the possibility of C–H activation through a radical process and indicate that the reaction proceeds through a two-electron process.

Table 2.3 Substrate scope

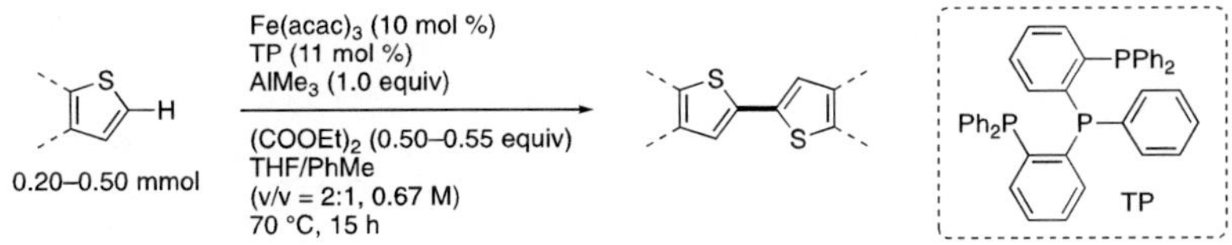

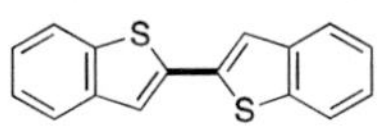

2
99% (41%)[a]

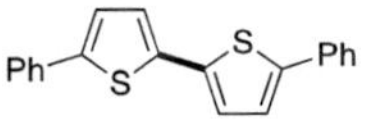

3
87%

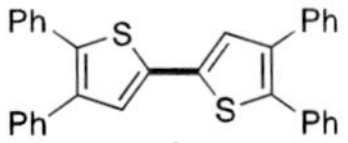

4
83%

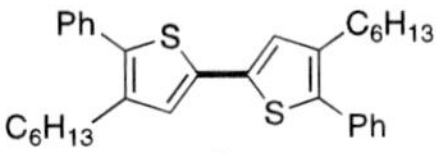

5
94%

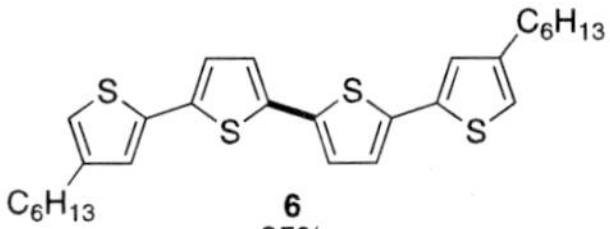

6
35%

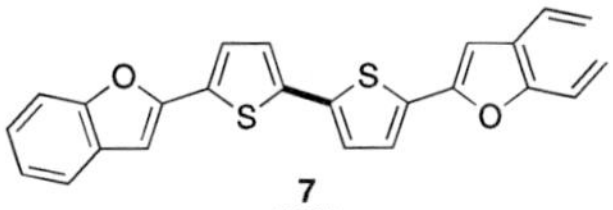

7
76%

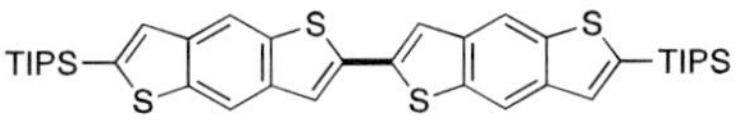

8
95%

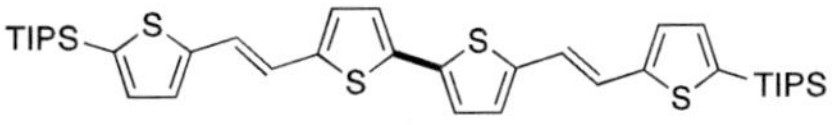

9
86%

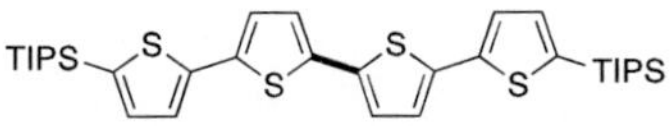

10
92%

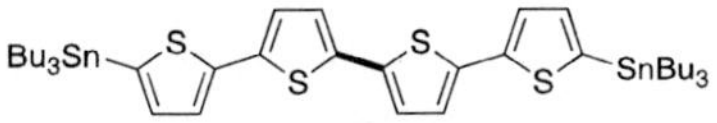

11
79%

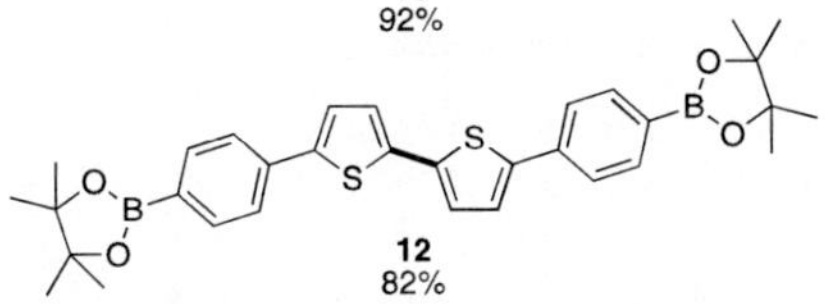

12
82%

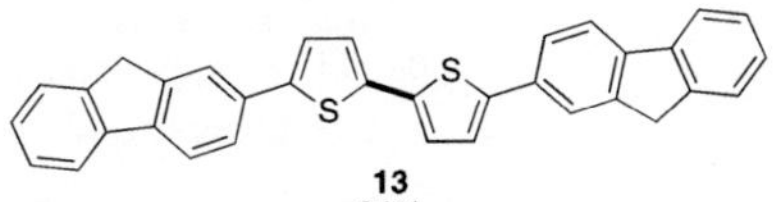

13
91%

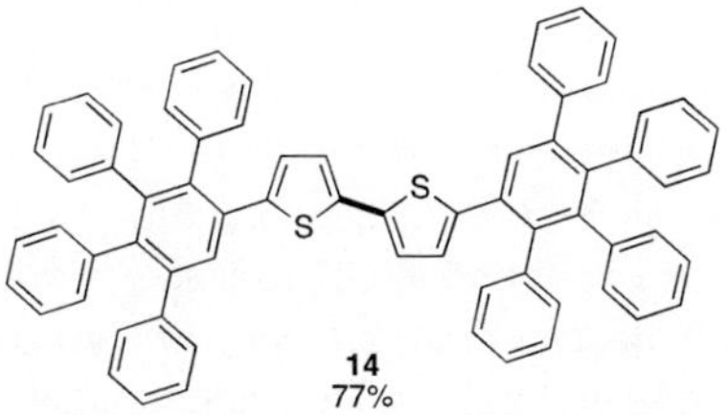

14
77%

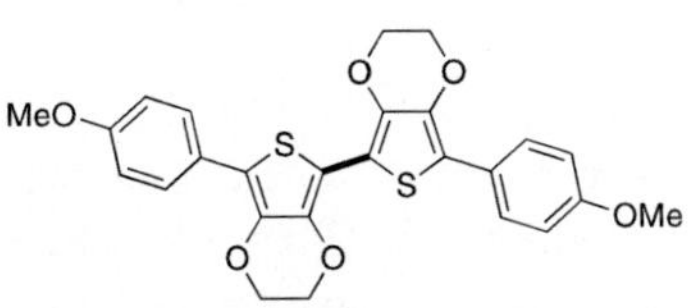

15
33%

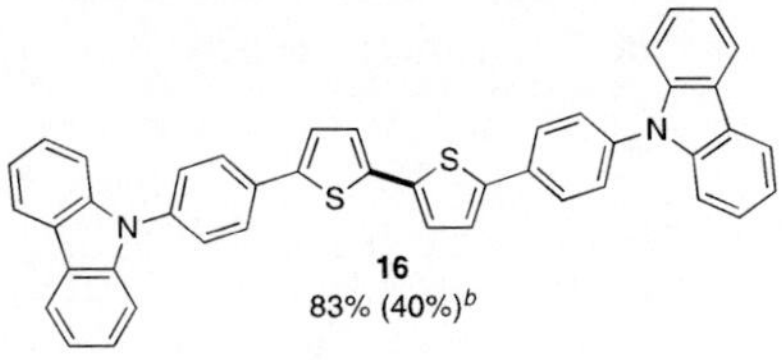

16
83% (40%)[b]

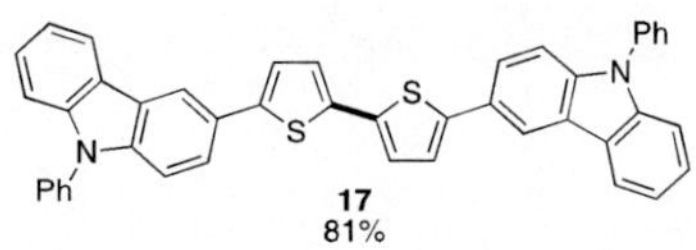

17
81%

(continued)

Table 2.3 (continued)

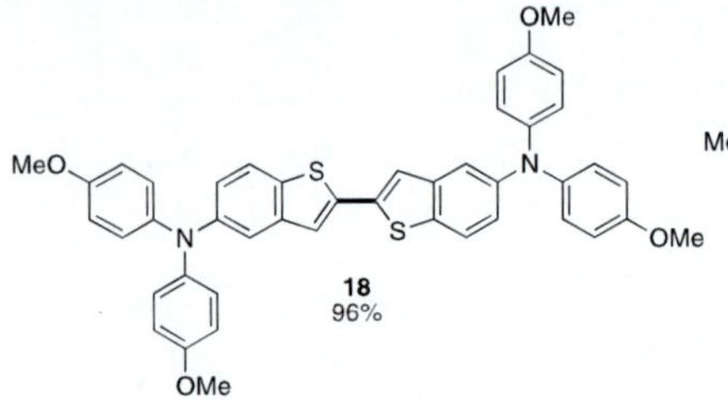

18
96%

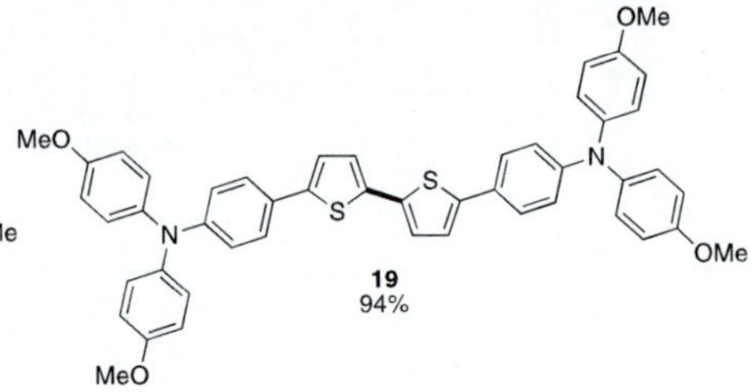

19
94%

All yields are isolated yields unless otherwise noted.
[a] Reported isolate yield under Mori's conditions: $PdCl_2(PhCN)_2$ (3.0 mol %), AgF (2.0 equiv), DMSO, 60 °C, 5 h.
[b] ^{1}H NMR yield under Mori's conditions.

Table 2.4 Unsuccessful substrates

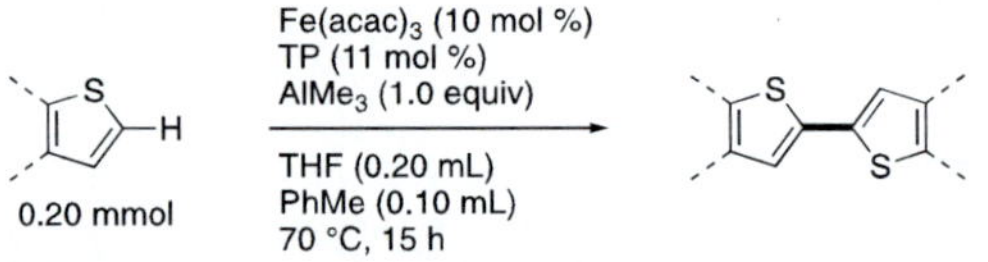

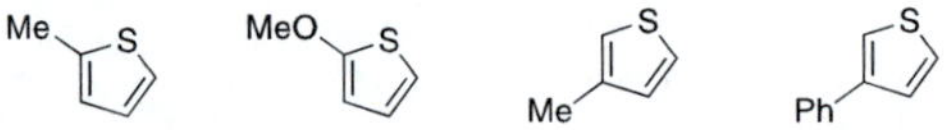

44%[a]	0%[b]	16%[a]	1%[b]

[a] Yields were determined by ^{1}H NMR using 1,1,2,2-tetrachloroethane as an internal standard.
[b] The yield was estmated by GC using tridecane as an internal standard.

2.9 The Role of $AlMe_3$ as a Base

To confirm the role of $AlMe_3$ as a base in iron-catalyzed thienyl C–H activation through a two-electron process, detection of CDH_3 was attempted using benzo[*b*]thiophene-2-*d* as a substrate. The reaction was conducted in a sealed Schlenk tube, and the gas was collected by the water displacement method and analyzed by ^{1}H NMR. A triplet peak of CDH_3 was observed, which provides evidence that the methyl group coming from $AlMe_3$ abstracts proton from the thiophene substrate (Scheme 2.13 and Fig. 2.3).

2.10 Stoichiometric Experiments

To obtain direct evidence that the C–H bond is activated by iron catalyst, stoichiometric experiments were conducted. When 1.0 equiv of benzo[*b*]thiophene was treated with 1.0 equiv of $AlMe_3$ without iron catalyst at 70 °C for 15 h and quenched with D_2O, no deuterium incorporation was observed (Scheme 2.14a). This indicates

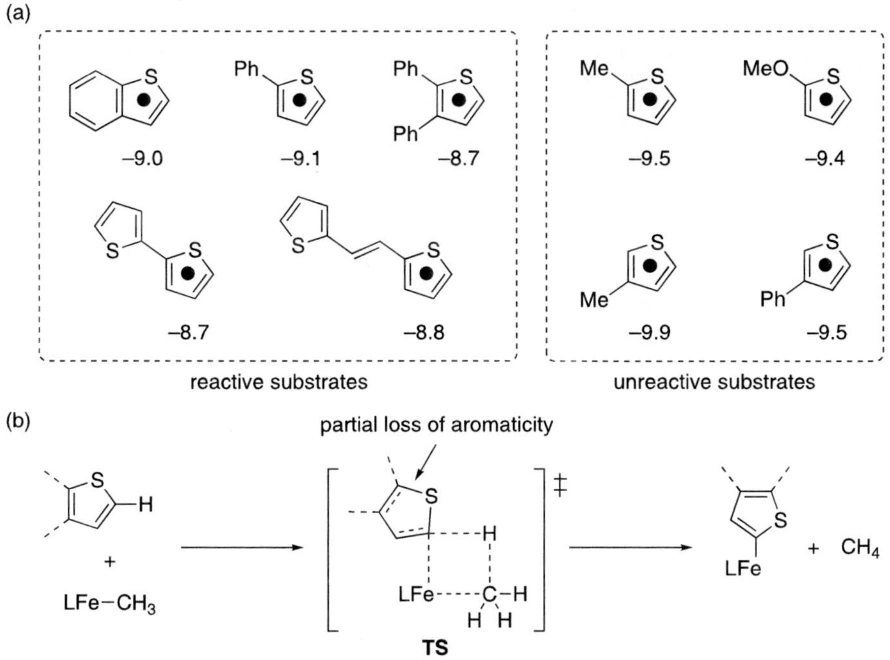

Fig. 2.2 Reactivity of π-extended and non-π-extended thiophenes. **a** NICS(1) values. **b** Partial loss of aromaticity in the transition state

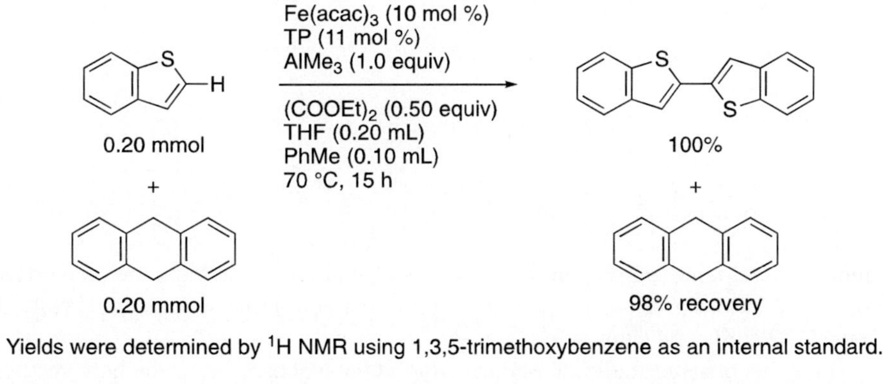

Scheme 2.12 Iron-catalyzed regioselective thienyl C–H/C–H homocoupling in the presence of a radical scavenger

that there is no alunimation to form a C–Al bond by deprotonation of a C–H bond by $AlMe_3$. On the other hand, when a similar experiment was performed in the presence of a stoichiometric amount of $Fe(acac)_3$/TP, 92% of deuterium incorporation on the starting material was observed, accompanying the formation of a small

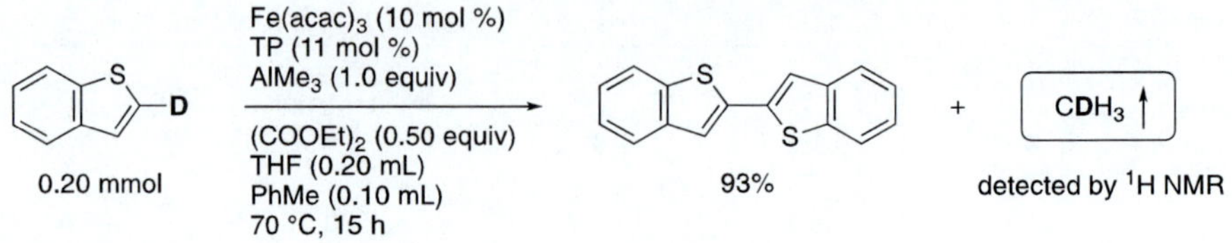

The yield was determined by ^{1}H NMR using 1,3,5-trimethoxybenzene as an internal standard.

Scheme 2.13 Detection of CDH_3

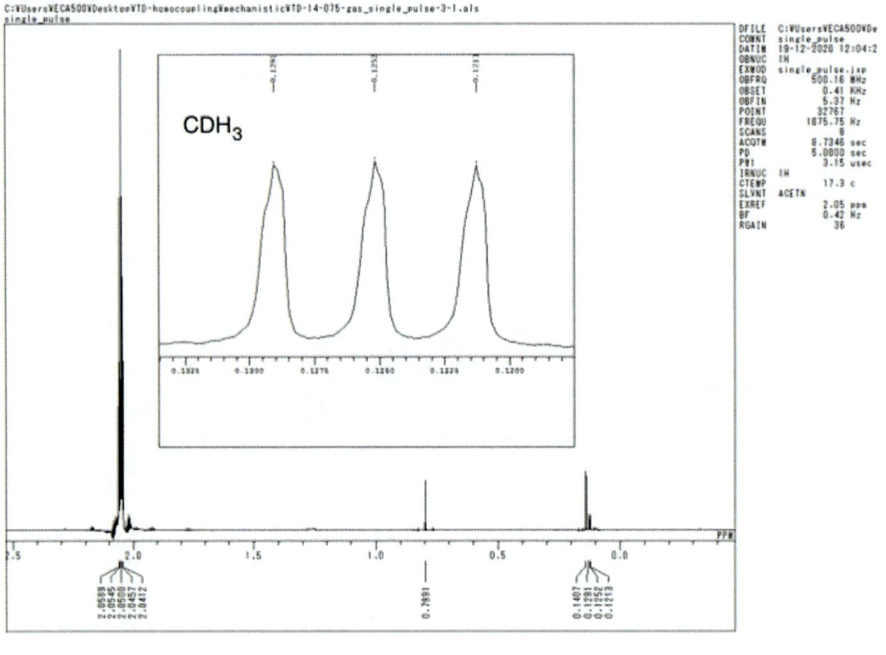

Fig. 2.3. ^{1}H NMR spectrum of CDH_3

amount of the homocoupling product and a trace amount of the methylation product (Scheme 2.14b). These results strongly suggest that Fe-thiophene species is involved in the C–H activation steps.

These mechanistic studies altogether support the mechanism in which the methyl group of $AlMe_3$ transfers to iron catalyst to form an Fe–Me species, which activates the C–H bond of a thiophene through σ-bond metathesis to generate an Fe–thiophene species by releasing methane (Scheme 2.15). Subsequent second C–H activation and reductive elimination steps will give the observed homocoupling product.

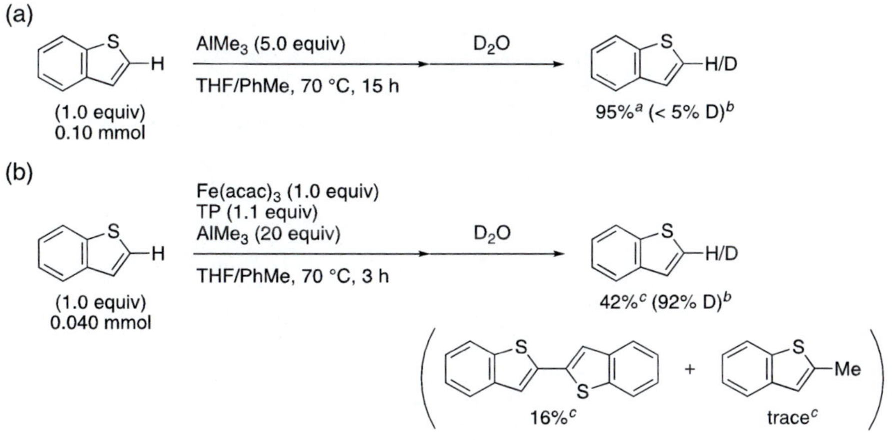

[a] The yield was determined by GC using tridecane as an internal standard.
[b] The deuterium incorporation ratio was determined by ^1H NMR.
[c] Yields were determined by ^1H NMR using 1,1,2,2-tetrachloroethane as an internal standard.

Scheme 2.14 Stoichiometric experiments

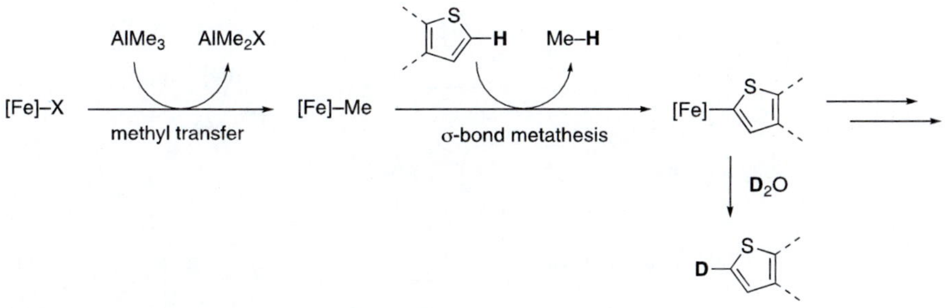

Scheme 2.15 Proposed mechanism of C–H activation

2.11 Kinetic Isotope Effect Experiments

To examine if either of the C–H activation steps is the turnover-limiting step, kinetic isotope effect experiments were conducted for two parallel reactions (Scheme 2.16). The initial reaction rates of benzo[*b*]thiophene and benzo[*b*]thiophene-2-*d* in a separate reaction vessels were measured by tracing the yields of the homocoupling product up to 16%. As a result, no difference of reaction rates was observed between two reactions. This suggests that neither of the C–H activation steps is the turnover-limiting step and the catalyst regeneration is the turnover-limiting step [20]. This is in line with the assumed catalyst regeneration mechanism where strongly binding bidentate enediolate has to undergo ligand exchange from Fe(III) to Al(III) (Scheme 2.17).

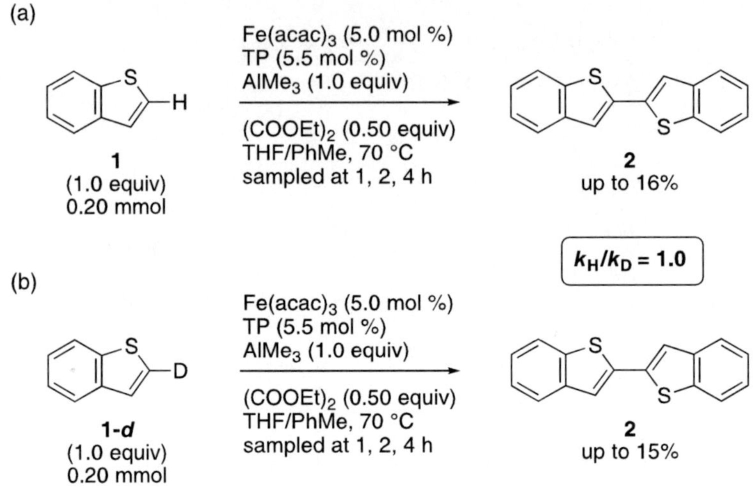

Yields were determined by ^{1}H NMR using 1,3,5-trimethoxybenzene as an internal standard.

Scheme 2.16 Kinetic isotope effect experiments for two parallel reactions

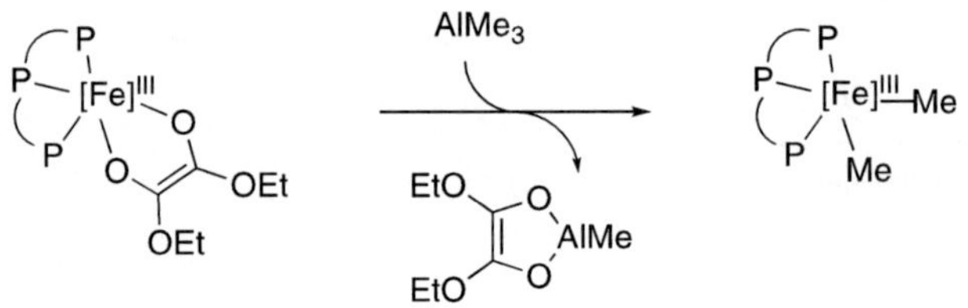

Scheme 2.17 Slow catalyst regeneration step

2.12 Conclusion

In conclusion, iron-catalyzed regioselective thienyl C–H/C–H homocoupling was developed using tridentate phosphine as a ligand, $AlMe_3$ as a base, and oxalate as a mild oxidant. Tridentate phosphine ligand was uniquely effective for this transformation and oxalate in combination with oxophilic $AlMe_3$ served as an effective but mild oxidant to oxidize Fe(I) species to regenerate Fe(III) catalyst. This reaction takes place exclusively at the C–H bond next to the sulfur atom of thienyl group and tolerates various kinds of redox-sensitive π motifs widely used in optoelectronic materials due to the mildness of the iron catalytic cycle created by the combination of low redox potential of Fe(III)/Fe(I) and mild oxalate oxidant. This work highlights the benefits of iron catalysis for the synthesis of π-conjugated dimeric and oligomeric compounds of importance in energy device applications. Further development of this homocoupling reaction to polymerization reaction will be discussed in the next chapter.

2.13 Experimental

Materials and methods

All air or moisture-sensitive reactions were performed in a dry reaction vessel under argon atmosphere. Air or moisture-sensitive liquids and solutions were transferred with syringe or Teflon cannula. The water content of solvents was confirmed to be less than 30 ppm by Karl Fischer titration performed with MKC-210 (Kyoto Electronics Manufacturing Co., Ltd.). Analytical thin-layer chromatography (TLC) was performed with a glass plate coated with 0.25 mm 230–400 mesh silica gel containing a fluorescent indicator. Organic solutions were evacuated with a diaphragm pump through a rotary evaporator. Flash column chromatography was performed as described by Still et al. [21] Preparative recycling gel permeation chromatography (GPC) was performed with LC-92XX II NEXT instrument (Japan Analytical Industry Co., Ltd.) equipped with JAIGEL-2 h polystyrene columns using chloroform as an eluent at the flow rate of 7.5 mL/min.

Gas chromatography (GC) was performed with GC-2014 instrument (Shimadzu Co.) equipped with an ULBON HR-1 (0.25 mm I.D. $\times$ 25 mL, 0.25 μm, Shinwa Chemical Industries, Ltd.) capillary column. Mass spectra (GC–MS) were taken with Parvum 2 instrument (Shimadzu Co.). High-resolution mass spectra (HRMS) were taken with LCMS-IT-TOF (Shimadzu Co.) using reserpine (MW 608.2734) as an internal standard. Melting points of solid compounds were measured on a Mel-Temp capillary melting-point apparatus and were uncorrected. Nuclear magnetic resonance (NMR) spectra were taken with ECZ-500 (JEOL, Ltd.) at room temperature unless otherwise noted and reported in parts per million (ppm). ^{1}H NMR spectra were internally referenced to tetramethylsilane (0.00 ppm), $CHCl_3$ (7.26 ppm), $CHDCl_2$ (5.32 ppm), C_2HDCl_4 (5.97 ppm), or $(CHD_2)(CD_3)SO$ (2.50 ppm). ^{13}C NMR spectra were internally referenced to tetramethylsilane (0.0 ppm), $CDCl_3$ (77.0 ppm), CD_2Cl_2 (53.8 ppm), $C_2D_2Cl_4$ (73.8 ppm), or $(CD_3)_2SO$ (39.5 ppm). ^{19}F NMR spectra were internally referenced to C_6F_6 (–164.9 ppm). ^{31}P NMR spectra were internally referenced to $(CH_3O)_3PO$ (2.1 ppm). ICP analysis was performed on Shimadzu ICPS-7510 equipment.

Unless otherwise noted, reagents were purchased from Tokyo Chemical Industry Co., Ltd., Sigma-Aldrich Co., LCC, FUJIFILM Wako Pure Chemical Co., and other commercial suppliers and were used as received. Anhydrous tetrahydrofuran and diethyl ether were purchased from KANTO Chemical Co., Inc. and purified prior to use by a solvent purification system (GlassContour) equipped with columns of activated alumina and supported copper catalyst [22]. $Fe(acac)_3$ (99.9% trace metal basis) was purchased from Sigma-Aldrich Co., LCC and used as received. Diethyl oxalate was purchased from Tokyo Chemical Industry Co., degassed by Freeze–Pump–Thaw cycling for three times, dried with molecular sieves 4A, and kept in a storage flask. Stating materials were synthesized according to the literature [23].

Preparation of ((Phenylphosphanediyl)bis(2,1-phenylene))bis(diphenylphosphane) (TP):

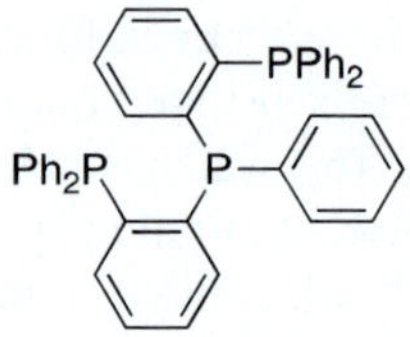

A hexane solution of BuLi (1.59 mol/L, 3.87 mL, 6.15 mmol) was added dropwise to a solution of (2-bromophenyl)diphenylphosphane (2.05 g, 6.00 mmol) in THF (12 mL) at –78 °C. After stirring for 1 h, dichloro(phenyl)phosphane (0.41 mL, 3.0 mmol) was added in one portion and the reaction mixture was gradually warmed to room temperature. The reaction mixture was stirred for 16 h and quenched with water (50 mL). The aqueous layer was extracted with dichloromethane (500 mL), and the combined organic layers were washed with brine and dried over Na_2SO_4. Most of the solvent was removed under reduced pressure and methanol (100 mL) was added under sonication. The white precipitate was collected by filtration and washed with methanol to afford the product as white solid (1.7 g, 91%). The compound data was in good agreement with the literature [6].

A representative procedure for the investigation of key reaction parameters of homocoupling (Table 2.2)

In an oven-dried Schlenk tube was added benzo[*b*]thiophene (27 mg, 0.20 mmol), TP (14 mg, 0.022 mmol), and a THF solution of $Fe(acac)_3$ (0.10 mol/L, 0.20 mL, 0.020 mmol). Then, a toluene solution of $AlMe_3$ (2.0 mol/L, 0.10 mL, 0.20 mmol) was added dropwise at room temperature and the reaction mixture was stirred for 5 min to give a clear dark reddish brown solution. Diethyl oxalate (14 μL, 0.10 mmol) was added and the reaction mixture was stirred at 70 °C for 15 h. The reaction mixture was cooled to rt, diluted with ethyl acetate (2 mL), and quenched carefully with methanol (0.1 mL). A saturated aqueous solution of potassium sodium tartrate (1 mL) was added, and the mixture was stirred vigorously until clear phase separation was observed. Tridecane (30 μL) was added as an internal standard and a portion of the organic layer was passed through a pad of Florisil and analyzed by GC. For ^{1}H NMR analysis, the crude mixture was extracted with chloroform and 1,3,5-trimethoxybenzene was added as an internal standard. A portion of the organic layer was passed through a pad of Florisil, and the solvent was removed under reduced pressure and the crude mixture was analyzed by ^{1}H NMR.

A general procedure for iron-catalyzed thienyl C–H/C–H homocoupling (Table 2.3)

In an oven-dried Schlenk tube was added a thiophene (0.40 mmol), TP (28 mg, 0.044 mmol), and a THF solution of $Fe(acac)_3$ (0.10 mol/L, 0.40 mL, 0.040 mmol). Then, a toluene solution of $AlMe_3$ (2.0 mol/L, 0.20 mL, 0.40 mmol) was added dropwise at rt, and the reaction mixture was stirred for 5 min to give a clear dark reddish brown solution. Diethyl oxalate (27 μL, 0.20 mmol) was added, and the reaction mixture was stirred at 70 °C for 15 h. The reaction mixture was cooled to rt, diluted with ethyl acetate (4 mL), and quenched carefully with methanol (0.2 mL). A saturated aqueous solution of potassium sodium tartrate (2 mL) was added, and the mixture was stirred vigorously until clear phase separation was observed. The aqueous layer was extracted with dichloromethane, and the combined organic layers were washed with brine and dried over Na_2SO_4. The solvent was removed under reduced pressure, and the crude product was purified by the indicated method below to afford the homocoupling product.

2,2′-Bibenzo[*b*]thiophene (2):

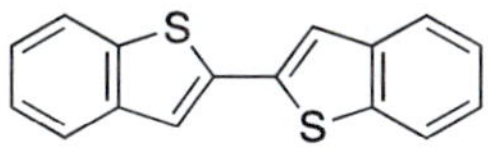

The title compound was obtained as white solid in 99% yield. The reaction was performed on a 0.40 mmol scale, and the crude product was purified by silica gel chromatography (dichloromethane only). The compound data was in good agreement with the literature [24].

^{1}H NMR (500 MHz, $CDCl_3$): δ 7.82 (d, J = 7.7 Hz, 2H), 7.77 (d, J = 7.7 Hz, 2H), 7.52 (s, 2H), 7.38–7.32 (m, 4H).

^{13}C NMR (125 MHz, $CDCl_3$): δ 140.2, 139.4, 137.2, 124.9, 124.8, 123.7, 122.2, 121.4.

5,5′-Diphenyl-2,2′-bithiophene (3):

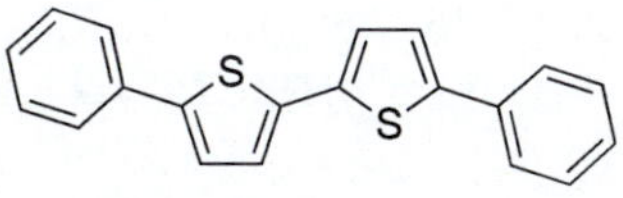

The title compound was obtained as yellow solid in 87% yield. The reaction was performed on a 0.40 mmol scale, and the crude product was purified by silica gel chromatography (dichloromethane only). The compound data was in good agreement with the literature [4].

^{1}H NMR (500 MHz, $CDCl_3$): δ 7.62–7.60 (m, 4H), 7.41–7.39 (m, 4H), 7.31–7.28 (m, 2H), 7.25 (d, J = 3.9 Hz, 2H), 7.18 (d, J = 3.9 Hz, 2H).

^{13}C NMR (125 MHz, $CDCl_3$): δ 143.1, 136.7, 134.0, 129.0, 127.6, 125.6, 124.5, 123.8.

4,4′,5,5′-Tetraphenyl-2,2′-bithiophene (4):

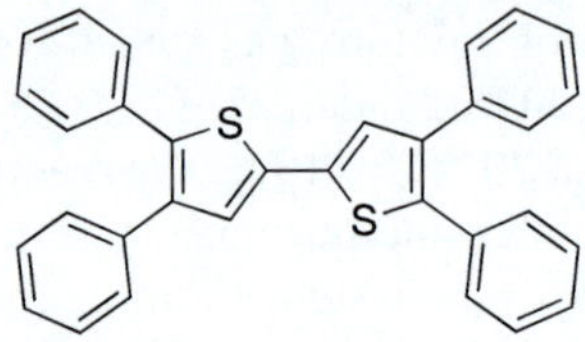

The title compound was obtained as yellow solid in 83% yield. The reaction was performed on a 0.20 mmol scale, and the crude product was purified by silica gel chromatography (hexane: dichloromethane = 19/1 to 9/1). The compound data was in good agreement with the literature [25].

^{1}H NMR (500 MHz, CD_2Cl_2): δ 7.34–7.27 (m, 22H).

^{13}C NMR (125 MHz, CD_2Cl_2): δ 139.3, 137.9, 136.6, 135.7, 134.2, 129.5, 129.4, 128.9, 128.8, 128.0, 128.0, 127.5, 127.5, 127.4, 127.3. (Multiple signals were observed because of slow rotation of bonds.)

4,4′-Dihexyl-5,5′-diphenyl-2,2′-bithiophene (5):

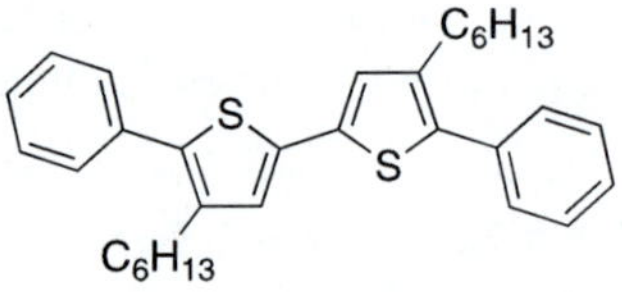

The title compound was obtained as yellow solid in 94% yield. The reaction was performed on a 0.40 mmol scale, and the crude product was purified by silica gel chromatography (dichloromethane only).

Melting point: 65–66 °C (dichloromethane).

^{1}H NMR (500 MHz, $CDCl_3$): δ 7.46–7.39 (m, 8H), 7.34–7.31 (m, 2H), 7.05 (s, 2H), 2.63 (t, $J = 7.8$ Hz, 4H), 1.64–1.61 (m, 4H), 1.34–1.25 (m, 12H), 0.86 (t, $J = 6.9$ Hz, 6H).

^{13}C NMR (125 MHz, $CDCl_3$): δ 139.5, 136.6, 135.5, 134.4, 129.2, 128.5, 127.3, 125.9, 31.6, 30.9, 29.2, 28.8, 22.6, 14.1.

HRMS (APCI+): *m/z* calcd for $C_{32}H_{38}S_2$ [M + H$^+$] 487.2488; found: 487.2498.

4,4‴-Dihexyl-2,2′:5′,2″:5″,2‴-quaterthiophene (6):

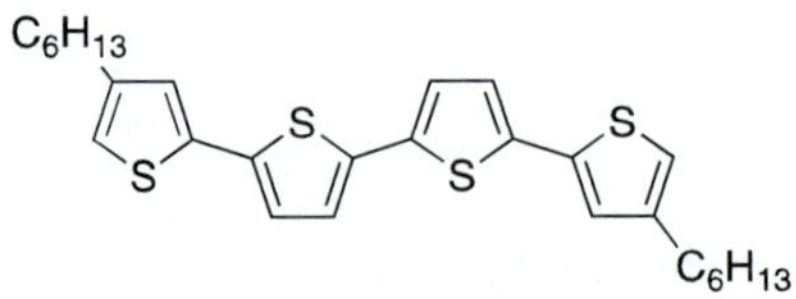

The title compound was obtained as orange solid in 35% yield. The reaction was performed on a 0.40 mmol scale, and the crude product was purified by silica gel chromatography (hexane only).

Melting point: 109–110 °C (dichloromethane).

^{1}H NMR (500 MHz, $CDCl_3$): δ 7.05–7.04 (m, 4H), 7.01 (d, $J = 1.3$ Hz, 2H), 6.81 (d, $J = 1.3$ Hz, 2H), 2.58 (t, $J = 7.8$ Hz, 4H), 1.64–1.55 (m, 4H), 1.37–1.30 (m, 12H), 0.90 (t, $J = 7.2$ Hz, 6H).

^{13}C NMR (125 MHz, $CDCl_3$): δ 144.2, 136.6, 136.6, 135.6, 125.1, 124.1, 124.0, 119.2, 31.7, 30.5, 30.4, 29.0, 22.6, 14.1.

HRMS (APCI+): *m/z* calcd for $C_{28}H_{34}S_4$ [M + H$^+$] 499.1616; found: 499.1630.

5,5′-Di(benzofuran-2-yl)-2,2′-bithiophene (7):

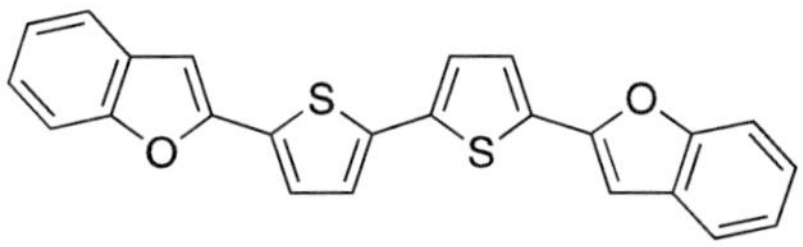

The title compound was obtained as orange solid in 76% yield. The reaction was performed on a 0.50 mmol scale using 0.55 equiv of diethyl oxalate and the pure product was obtained by filtration of the crude reaction mixture and by washing the solid with minimal amounts of THF, an aqueous solution of HCl (1 M), and acetone.

Melting point: 267–268 °C (THF/toluene).

^{1}H NMR (500 MHz, $CDCl_3$): δ 7.57 (d, $J = 7.8$ Hz, 2H), 7.51 (d, $J = 8.1$ Hz, 2H), 7.42 (d, $J = 3.7$ Hz, 2H), 7.31–7.28 (m, 2H), 7.25–7.22 (m, 4H), 6.90 (s, 2H).

^{13}C NMR (125 MHz, $CDCl_3$): δ 154.6, 150.6, 137.1, 132.2, 129.0, 125.4, 124.7, 124.5, 123.2, 120.8, 111.1, 101.5.

HRMS (APCI+): *m/z* calcd for $C_{24}H_{14}O_2S_2$ [M + H$^+$] 399.0508; found: 399.0503.

6,6′-Bis(triisopropylsilyl)-2,2′-bibenzo[1,2-*b*:4,5-*b*′]dithiophene (8):

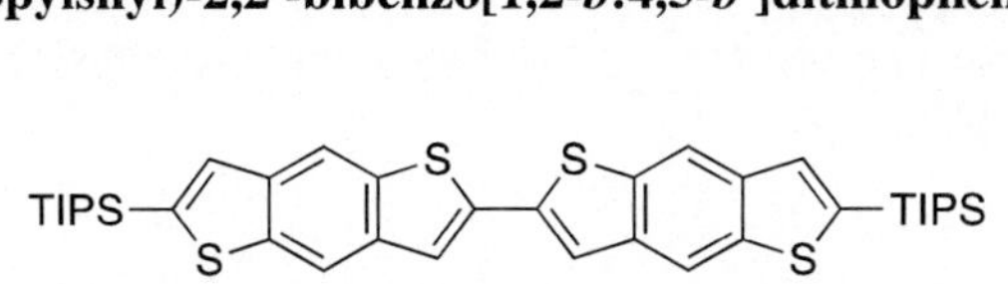

The title compound was obtained as yellow solid in 95% yield. The reaction was performed on a 0.20 mmol scale, and the crude product was purified by silica gel chromatography (hexane: chloroform = 5:1).

Melting point: 239–241 °C (chloroform).

^{1}H NMR (500 MHz, $CDCl_3$): δ 8.24 (s, 2H), 8.23 (s, 2H), 7.55 (s, 2H), 7.51 (s, 2H), 1.44 (sep, J = 7.6 Hz, 6H), 7.6 (d, J = 7.6 Hz, 36H).

^{13}C NMR (125 MHz, $CDCl_3$): δ 141.3, 139.4, 138.6, 137.9, 137.8, 136.5, 131.5, 120.7, 116.3, 116.1, 18.6, 11.8.

HRMS (APCI+): m/z calcd for $C_{38}H_{50}O_4Si_2$ [M + H^+] 691.2407; found: 691.2392.

5,5′-Bis((*E*)-2-(5-(triisopropylsilyl)thiophen-2-yl)vinyl)-2,2′-bithiophene (9):

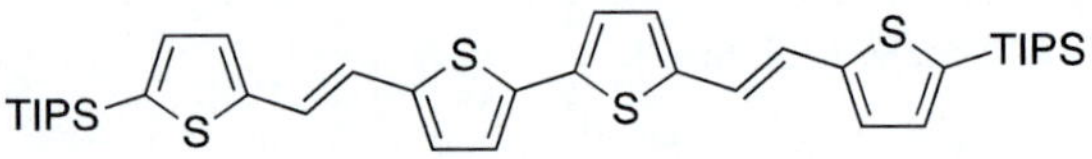

The title compound was obtained as orange solid in 86% yield. The reaction was performed on a 0.20 mmol scale, and the crude product was purified by silica gel chromatography (hexane: chloroform = 5:1).

Melting point: 161–162 °C (chloroform).

^{1}H NMR (500 MHz, $CDCl_3$): δ 7.15 (d, J = 3.5 Hz, 2H), 7.12 (d, J = 3.5, 2H), 7.07–7.00 (m, 6H), 6.92 (d, J = 3.7 Hz, 2H), 1.34 (sep, J = 7.5 Hz, 6H), 1.12 (d, J = 7.5 Hz, 36H).

^{13}C NMR (125 MHz, $CDCl_3$): δ 147.2, 141.7, 136.4, 136.0, 134.4, 127.3, 127.1, 124.2, 121.5, 121.3, 18.6, 11.8.

HRMS (APCI+): m/z calcd for $C_{38}H_{54}S_4Si_2$ [M + H^+] 695.2720; found: 695.2733.

5,5‴-Bis(triisopropylsilyl)-2,2′:5′,2″:5″,2‴-quaterthiophene (10):

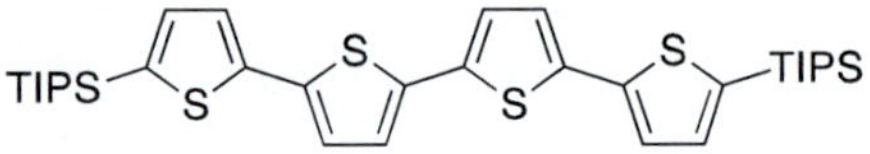

The title compound was obtained as yellow solid in 92% yield. The reaction was performed on a 0.20 mmol scale, and the crude product was purified by silica gel chromatography (hexane: chloroform = 5:1). The compound data was in good agreement with the literature [26].

^{1}H NMR (500 MHz, C_6D_6): δ 6.80 (d, J = 3.4 Hz, 2H), 6.64 (d, J = 3.4 Hz, 2H), 6.52 (d, J = 3.7 Hz, 2H), 6.43 (d, J = 3.7 Hz, 2H), 0.87 (sep, J = 7.4 Hz, 6H), 0.73 (d, J = 7.4 Hz, 36H).

^{13}C NMR (125 MHz, C_6D_6): δ 142.7, 137.0, 136.7, 136.4, 133.9, 125.3, 125.0, 124.7, 18.7, 12.1.

5,5‴-Bis(tributylstannyl)-2,2′:5′,2″:5″,2‴-quaterthiophene (11):

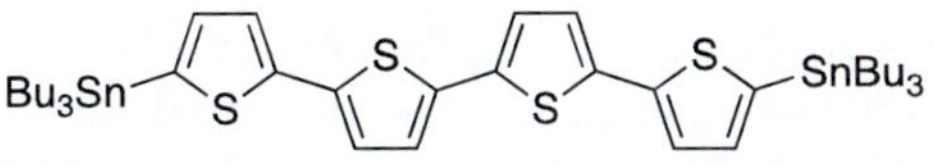

The title compound was obtained as orange oil in 79% yield. The reaction was performed on a 0.20 mmol scale, and the crude product was passed through a pad of Florisil and purified by gel permeation chromatography (toluene). Purification by silica gel chromatography failed because of the destannylation of the product.

^{1}H NMR (500 MHz, C_6D_6): δ 6.91 (d, $J = 3.3$ Hz, 2H), 6.69 (d, $J = 3.3$ Hz, 2H), 6.53 (d, $J = 3.7$ Hz, 2H), 6.42 (d, $J = 3.7$ Hz, 2H), 1.26–1.20 (m, 12H), 1.00–0.93 (m, 12H), 0.74–0.71 (m, 12H), 0.52 (t, $J = 7.2$ Hz, 18H).

^{13}C NMR (125 MHz, $CDCl_3$): δ 142.4, 137.1, 136.4, 136.2, 135.6, 124.9, 124.1 (two signals overlapped), 28.9, 27.3, 13.7, 10.9.

HRMS (APCI+): *m/z* calcd for $C_{40}H_{62}S_4Sn_2$ [M + H$^+$] 909.1853; found: 909.1840.

5,5′-Bis(4-(4,4,5,5-tetramethyl-1,3,2-dioxaborolan-2-yl)phenyl)-2,2′-bithiophene (12):

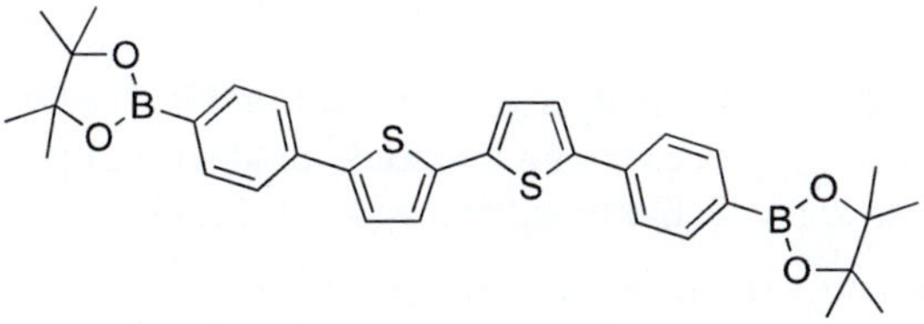

The title compound was obtained as yellow solid in 82% yield. The reaction was performed on a 0.20 mmol scale using 0.55 equiv of diethyl oxalate, and the pure product was obtained by filtration of the crude reaction mixture and by washing the solid with minimal amounts of THF, an aqueous solution of HCl (1 M), and acetone. The filtrate was collected and further purified by silica gel chromatography (dichloromethane only). The compound data was in good agreement with the literature [27].

^{1}H NMR (500 MHz, $CDCl_3$): δ 7.83–7.81 (m, 4H), 7.62–7.60 (m, 4H), 7.31 (d, $J = 3.9$ Hz, 2H), 7.19 (d, $J = 3.9$ Hz, 2H), 1.36 (s, 24H).

^{13}C NMR (125 MHz, $CDCl_3$): δ 143.0, 137.1, 136.4, 135.4, 128.3 (determined by HMBC), 124.6, 124.6, 124.4, 83.9, 24.9.

5,5′-Di(9*H*-fluoren-2-yl)-2,2′-bithiophene (13):

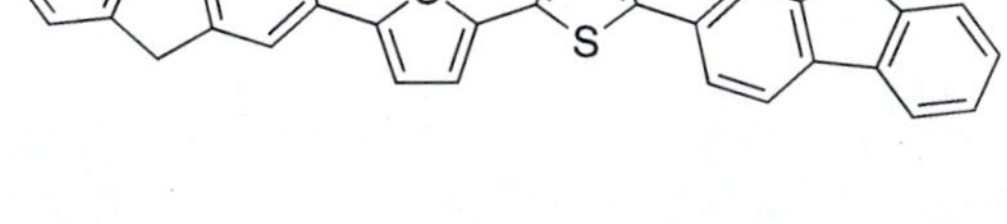

The title compound was obtained as yellow solid in 91% yield. The reaction was performed on a 0.50 mmol scale using 0.55 equiv of diethyl oxalate and the pure product was obtained by filtration of the crude reaction mixture and by washing the solid with minimal amounts of THF, an aqueous solution of HCl (1 M), and acetone. The compound data was in good agreement with the literature [4].

^{1}H NMR (500 MHz, $C_2D_2Cl_4$, 125 °C): δ 7.85–7.78 (br, 6H), 7.71–7.65 (br, 2H), 7.61–7.55 (br, 2H), 7.44–7.22 (br, 8H), 3.98 (br s, 4H).

5,5′-Bis(3′,6′-diphenyl-[1,1′:2′,1″-terphenyl]-4′-yl)-2,2′-bithiophene (14):

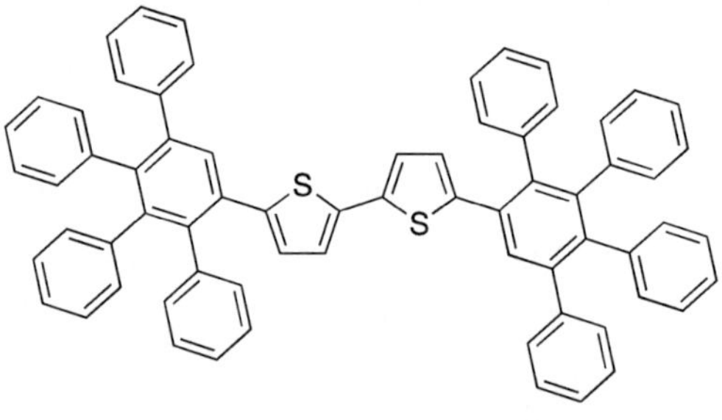

The title compound was obtained as yellow solid in 77% yield. The reaction was performed on a 0.20 mmol scale, and the crude product was purified by silica gel chromatography (hexane: dichloromethane = 4:1 to 2:1).

Melting point: 363–365 °C (dichloromethane).

^{1}H NMR (500 MHz, $CDCl_3$): δ 7.70 (s, 2H), 7.18–7.15 (m, 10H), 7.07–7.05 (m, 6H), 7.00–6.98 (m, 4H), 6.92–6.90 (m, 6H), 6.86–6.80 (m, 10H), 6.78–6.75 (m, 6H), 6.44 (d, J = 3.8 Hz, 2H).

^{13}C NMR (125 MHz, $CDCl_3$): δ 142.3, 142.0, 141.3, 141.0, 140.0, 139.7, 139.7, 139.6, 138.9, 137.4, 132.8, 131.3, 131.2, 131.1, 130.7, 129.9, 127.9, 127.6, 127.5, 126.9, 126.6, 126.4, 126.4, 125.7, 125.4, 123.2.

HRMS (APCI+): *m/z* calcd for $C_{68}H_{46}S_2$ [M + H$^+$] 927.3114; found: 927.3106.

7,7′-Bis(4-methoxyphenyl)-2,2′,3,3′-tetrahydro-5,5′-bithieno[3,4-*b*][1,4]dioxine (15):

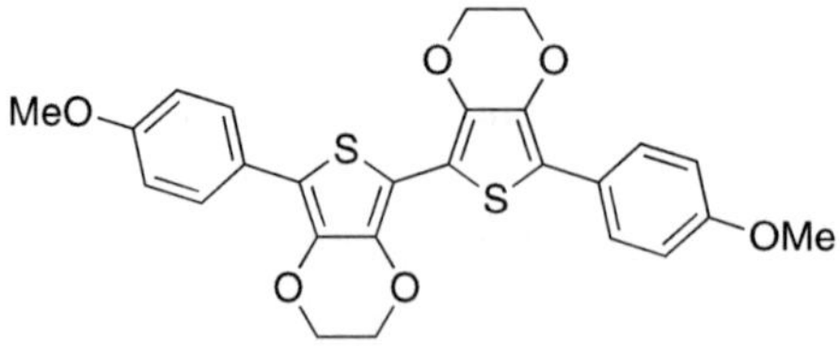

The title compound was obtained as yellow solid in 33% yield. The reaction was performed on a 0.40 mmol scale, and the pure product was obtained by filtration of the crude reaction mixture and by washing the solid with minimal amounts of THF, an aqueous solution of HCl (1 M), and acetone.

Melting point: 310 °C, decomp.

^{1}H NMR (500 MHz, $(CD_3)_2SO$): δ 7.54–7.51 (m, 4H), 6.92–6.89 (m, 4H), 4.35–4.30 (m, 8H), 3.71 (s, 6H).

^{13}C NMR (125 MHz, $(CD_3)_2SO$): δ 157.9, 137.4, 137.0, 126.6, 125.1, 114.3, 113.5, 105.9, 64.9, 64.6, 55.1.

HRMS (APCI+): *m/z* calcd for $C_{26}H_{22}O_6S_2$ [M + H$^+$] 495.0931; found: 495.0924.

5,5′-Bis(4-(9*H*-carbazol-9-yl)phenyl)-2,2′-bithiophene (16):

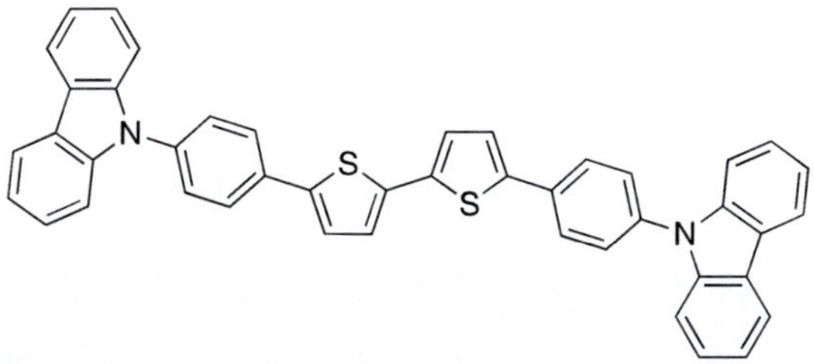

The title compound was obtained as yellow solid in 83% yield. The reaction was performed on a 0.50 mmol scale using 0.55 equiv of diethyl oxalate, and the pure product was obtained by filtration of the crude reaction mixture and by washing the solid with minimal amounts of THF, an aqueous solution of HCl (1 M), and acetone. This compound is known [28].

^{1}H NMR (500 MHz, $CDCl_3$): δ 8.17 (d, J = 7.7 Hz, 4H), 7.87–7.85 (m, 4H), 7.63–7.62 (m, 4H), 7.49–7.43 (m, 8H), 7.38 (d, J = 4.0 Hz, 2H), 7.33–7.28 (m, 6H).

^{13}C NMR (125 MHz, $CDCl_3$): δ 142.3, 140.7, 137.1, 137.0, 133.0, 127.5, 126.9, 126.0, 124.9, 124.4, 123.5, 120.4, 120.1, 109.8.

5,5′-Bis(9-phenyl-9*H*-carbazol-3-yl)-2,2′-bithiophene (17):

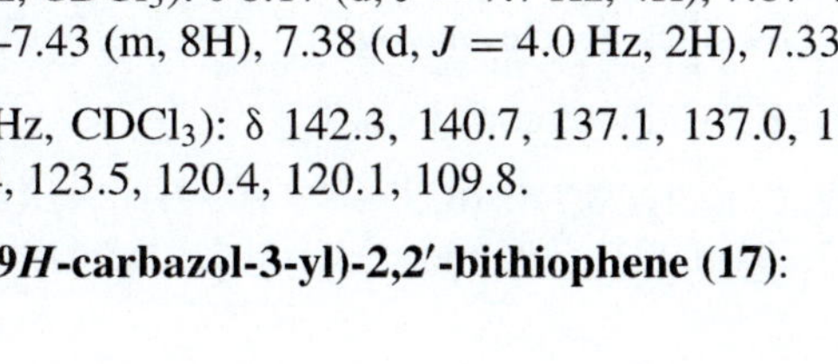

The title compound was obtained as yellow solid in 81% yield. The reaction was performed on a 0.50 mmol scale using 0.55 equiv of diethyl oxalate and the pure product was obtained by filtration of the crude reaction mixture and by washing the solid with minimal amounts of THF, an aqueous solution of HCl (1 M), and acetone. The filtrate was collected and further purified by silica gel chromatography (dichloromethane only). The compound data was in good agreement with the literature [29].

^{1}H NMR (500 MHz, $CDCl_3$): δ 8.38 (d, J = 1.5 Hz, 2H), 8.20 (d, J = 7.7 Hz, 2H), 7.69 (dd, J = 8.6, 1.7 Hz, 2H), 7.65–7.58 (m, 8H), 7.51–7.41 (m, 8H), 7.34–7.30 (m, 4H), 7.24 (d, J = 3.7 Hz, 2H).

^{13}C NMR (125 MHz, $CDCl_3$): δ 144.1, 141.4, 140.4, 137.4, 136.0, 130.0, 127.6, 127.0, 126.4, 126.3, 124.3, 124.2, 123.9, 123.2, 122.9, 120.5, 120.2, 117.4, 110.2, 110.0.

***N*5,*N*5,*N*5′,*N*5′-tetrakis(4-methoxyphenyl)-[2,2′-bibenzo[*b*]thiophene]-5,5′-diamine (18)**:

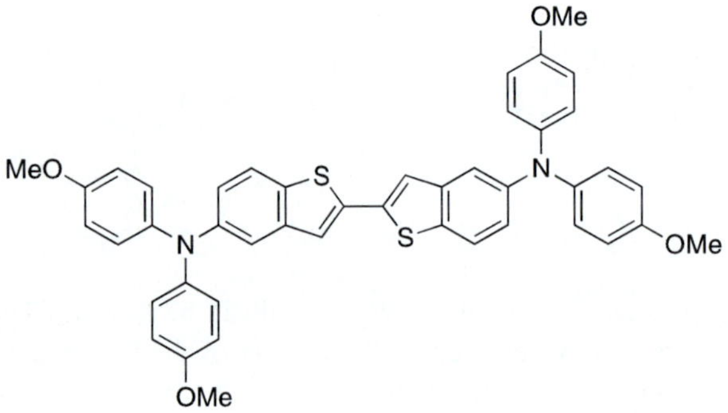

The title compound was obtained as yellow solid in 96% yield. The reaction was performed on a 0.20 mmol scale using 0.55 equiv of diethyl oxalate, and the crude product was purified by silica gel chromatography (dichloromethane only).

Melting point: 221–223 °C (dichloromethane).

^{1}H NMR (500 MHz, CD_2Cl_2): δ 7.58 (d, J = 8.9 Hz, 2H), 7.28 (s, 2H), 7.25 (d, J = 2.3 Hz, 2H), 7.05–7.03 (m, 8H), 7.01 (dd, J = 8.9, 2.3 Hz, 2H), 6.84–6.82 (m, 8H), 3.78 (s, 12H).

^{13}C NMR (125 MHz, CD_2Cl_2): δ 156.1, 147.1, 141.7, 141.6, 138.2, 132.4, 126.5, 122.7, 121.2, 121.2, 115.8, 115.0, 55.8.

HRMS (APCI+): m/z calcd for $C_{44}H_{36}N_2O_4S_2$ [$M + H^+$] 721.2189; found: 721.2190.

4,4′-([2,2′-Bithiophene]-5,5′-diyl)bis(*N*,*N*-bis(4-methoxyphenyl)aniline) (19):

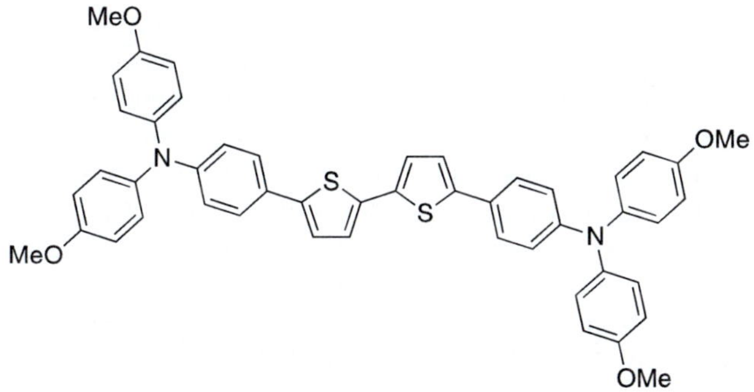

The title compound was obtained as orange solid in 94% yield. The reaction was performed on a 0.20 mmol scale, and the crude product was purified by silica gel chromatography (hexane: dichloromethane = 1:1). The compound data was in good agreement with the literature [17].

^{1}H NMR (500 MHz, CD_2Cl_2): δ 7.33–7.30 (m, 4H), 7.03–7.02 (m, 4H), 7.00–6.97 (m, 8H), 6.81–6.75 (m, 12H), 3.71 (s, 12H).

^{13}C NMR (125 MHz, CD_2Cl_2): δ 156.6, 148.8, 143.4, 140.8, 135.7, 127.2, 126.4, 126.1, 124.5, 122.6, 120.4, 115.0, 55.8.

Procedure for control experiment (Table 2.3, compound 16)

Mori's condition [3] was used for the synthesis of compound **16**. In an oven-dried Schlenk tube was added $PdCl_2(PhCN)_2$ (2.3 mg, 0.0060 mmol), DMSO (1.2 mL), and 9-(4-(thiophen-2-yl)phenyl)-9*H*-carbazole (65 mg, 0.20 mmol). To the resulting reaction mixture, silver(I) fluoride (51 mg, 0.40 mmol) was added and heated at 60 °C for 5 h. Then, the mixture was cooled to rt, diluted with chloroform (50 mL), and washed with water. 1,3,5-trimethoxybenzene was added as an internal standard, and a portion of the organic phase was passed through a pad for Florisil. The solvent was removed under reduced pressure, and the crude product was analyzed by ^{1}H NMR. The yield of **16** was determined to be 40%.

Procedure for detection of CDH_3 (Scheme 2.13)

In an oven-dried Schlenk tube was added benzo[*b*]thiophene-2-*d* (27 mg, 0.20 mmol), TP (14 mg, 0.022 mmol), and a THF solution of $Fe(acac)_3$ (0.10 mol/L, 0.20 mL, 0.020 mmol). Then, a toluene solution of $AlMe_3$ (2.0 mol/L, 0.10 mL, 0.20 mmol) was added dropwise at rt and the reaction mixture was stirred for 5 min to give a clear dark reddish brown solution. Diethyl oxalate (14 μL, 0.10 mmol) was added, and the reaction mixture was stirred at 70 °C for 15 h. After cooling to rt, the Schlenk tube was connected with a needle, and the gas was collected by the water displacement method by carefully opening the J. Young cap. 3 mL of the collected gas was taken by a gas tight syringe and was transferred to an NMR tube fitted with a septum and a short needle by the upward displacement method. The gas tight syringe and the short needle were removed quickly, and acetone-*d*6 was injected without releasing the pressure.

The NMR tube was shaken vigorously for 30 s, and the obtained sample was analyzed by ^{1}H NMR within 5 min. The triplet peak of CDH_3 was observed. The spectroscopic data was in good agreement with the literature [30]. The reaction mixture was diluted with ethyl acetate (2 mL) and quenched carefully with methanol (0.1 mL). A saturated aqueous solution of potassium sodium tartrate (1 mL) was added, and the mixture was stirred vigorously until clear phase separation was observed. The crude mixture was extracted with chloroform, and 1,3,5-trimethoxybenzene was added as an internal standard. A portion of the organic layer was passed through a pad of Florisil and the solvent was removed under reduced pressure and the crude mixture was analyzed by ^{1}H NMR. The yield of **2** was determined to be 93%.

Reaction of benzo[*b*]thiophene with $AlMe_3$ (Scheme 2.14a)

A mixture of benzo[*b*]thiophene (13 mg, 0.10 mmol) and $AlMe_3$ (2.0 mol/L, 0.25 mL, 0.50 mmol) in THF (0.25 mL) was stirred at 70 °C for 15 h. The reaction mixture was quenched carefully with D_2O (1.0 mL) and stirred at 70 °C for another 1 h. The reaction mixture was cooled to room temperature, and a saturated aqueous solution of potassium sodium tartrate (2.5 mL) was added. The mixture was stirred vigorously until clear phase separation was observed. The aqueous layer was extracted with diethyl ether (2 mL × 3), and the combined organic layers were passed through a pad of Florisil. The yield of recovered benzo[*b*]thiophene was determined by GC using tridecane as an internal standard, and the deuterium incorporation ratio was determined by ^{1}H NMR.

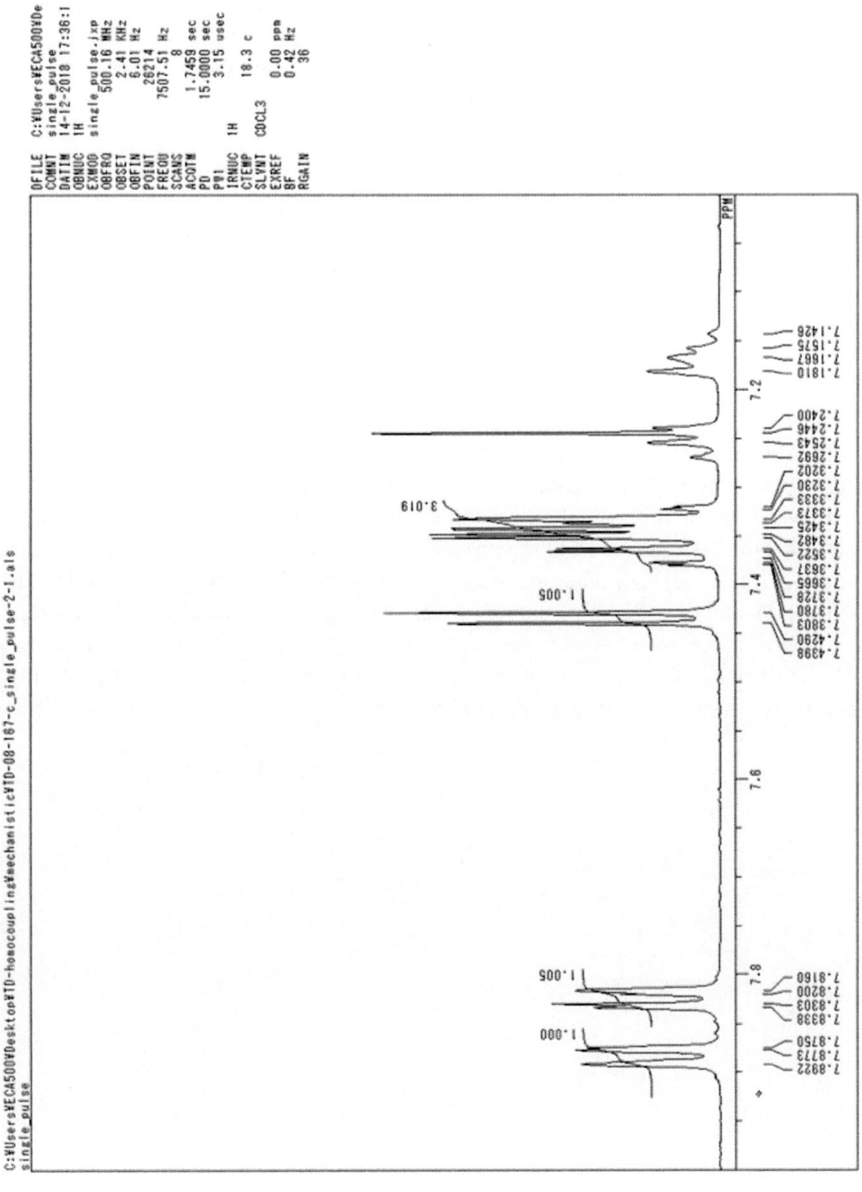

Procedure for stoichiometric experiment (Scheme 2.14b)

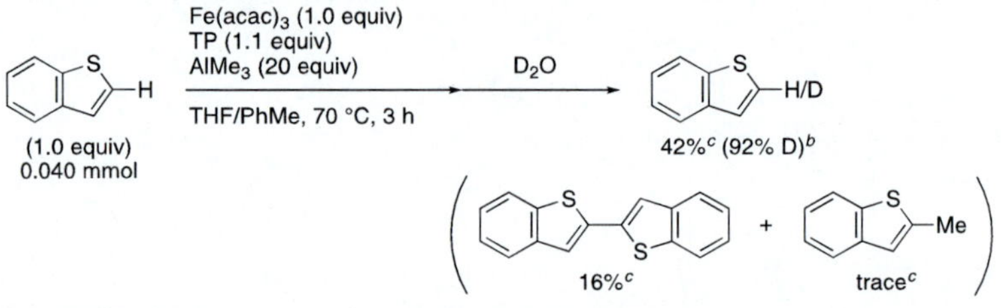

[a] The yield was determined by GC using tridecane as an internal standard.
[b] The deuterium incorporation ratio was determined by ^{1}H NMR.
[c] Yields were determined by ^{1}H NMR using 1,1,2,2-tetrachloroethane as an internal standard.

In an oven-dried Schlenk tube was added benzo[*b*]thiophene (5.4 mg, 0.040 mmol), TP (28 mg, 0.044 mmol), and a THF solution of $Fe(acac)_3$ (0.10 mol/L, 0.40 mL, 0.040 mmol). Then, a toluene solution of $AlMe_3$ (2.0 mol/L, 0.40 mL, 0.80 mmol) was added dropwise at room temperature, and the reaction mixture was stirred for 5 min to give a clear dark reddish brown solution. The reaction mixture was stirred at 70 °C for 3 h, cooled to room temperature, and quenched carefully with D_2O (1.0 mL). After stirring for 1 h, a saturated aqueous solution of potassium sodium tartrate (1.0 mL) was added, and the mixture was stirred vigorously until clear phase separation was observed. The aqueous layer was extracted with dichloromethane (20 mL × 3), and the combined organic layers were passed through a pad of Florisil. The solvent was removed under reduced pressure, and the crude mixture was analyzed by ^{1}H NMR using 1,1,2,2-tetrachloroethane as an internal standard. The deuterium incorporation ratio was also determined by ^{1}H NMR.

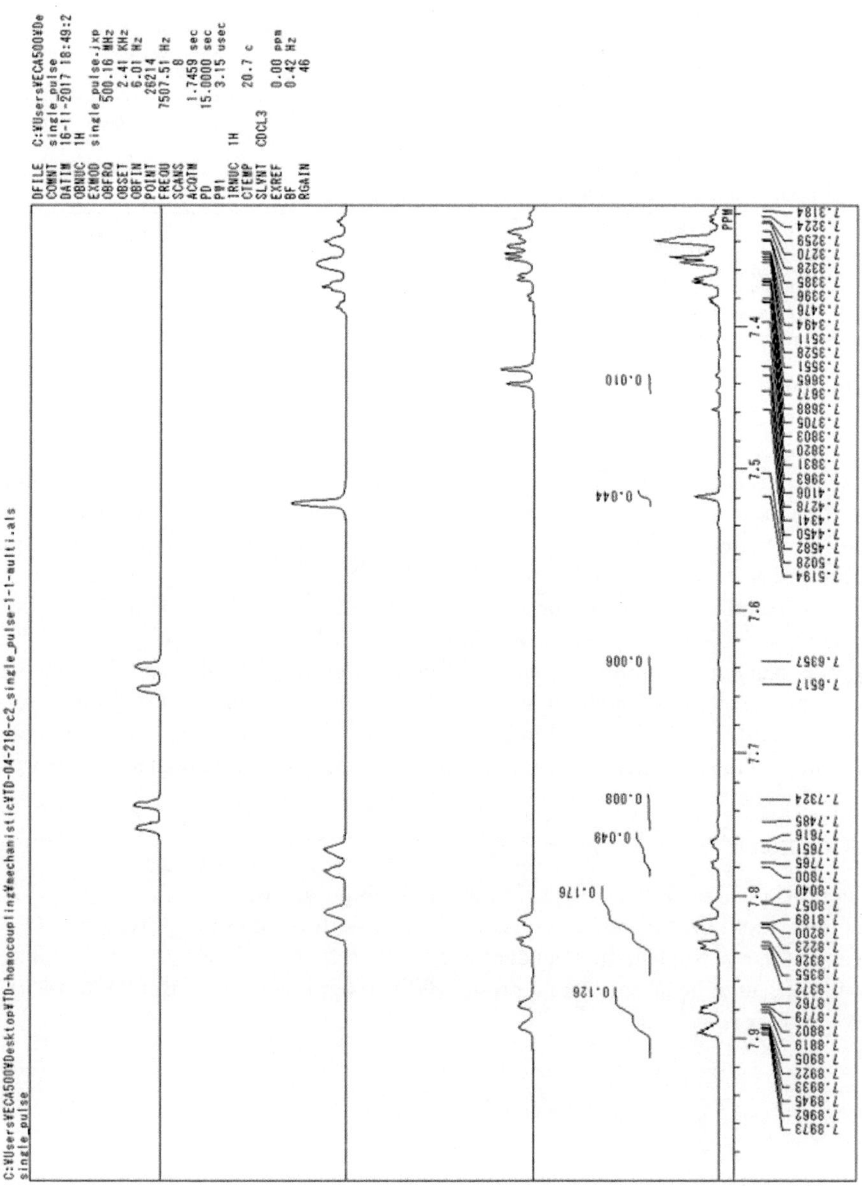

Kinetic isotope effect experiments for two parallel reactions (Scheme 2.16)

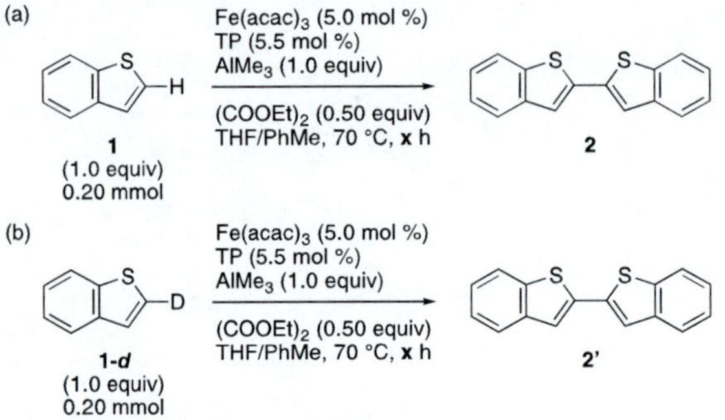

x	yield (%) 2	2'	k_H/k_D
1	9.1	9.0	1.0
2	12	12	1.0
4	16	15	1.0

$k_H/k_D = 1.0$

Yields were determined by ^{1}H NMR using 1,3,5-trimethoxybenzene as an internal standard.

In an oven-dried Schlenk tube was added benzo[*b*]thiophene (27 mg, 0.20 mmol), TP (6.9 mg, 0.011 mmol), a THF solution of $Fe(acac)_3$ and 1,3,5-trimethoxybenzene (0.10 mol/L, 0.10 mL, 0.010 mmol each), and THF (1.0 mL). Then, a toluene solution of $AlMe_3$ (2.0 mol/L, 0.10 mL, 0.20 mmol) was added dropwise at room temperature and the reaction mixture was stirred for 5 min to give a clear dark reddish brown solution. Diethyl oxalate (14 μL, 0.20 mmol) was added, and the reaction mixture was stirred at 70 °C and sampled after 1, 2, and 4 h. Exactly the same experiment using the same solutions of reagents and the same oil bath was conducted at the same time for benzo[*b*]thiophene-2-*d* (27 mg, 0.20 mmol). Each sample was quenched with a saturated aqueous solution of potassium sodium tartrate (0.5 mL), and the aqueous layer was extracted with dichloromethane (0.5 mL × 3). The combined organic layers were passed through a pad of Florisil and the solvent was removed under reduced pressure. All the samples were analyzed by ^{1}H NMR using the singlet aromatic peak of 1,3,5-trimethoxybenzene as an internal standard.

Calculation of NICS(1) values (Fig. 2.2a)

Theoretical calculations were performed using Gaussian 09 software package [31]. Geometry optimizations of all compounds were performed at the B3LYP/6-31G(d) level of theory. Nucleus-independent chemical shifts (NICS) [18] were calculated using the gauge invariant atomic orbital (GIAO) approach at the B3LYP/6-31G(d) level of theory.

Cartesian coordinates of the optimized structures (Bq stands for ghost atom):

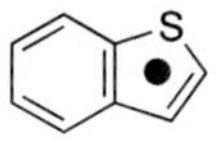

0 1

C	0.06190302	-0.55391131	0.00006416
C	-0.11104371	0.85304021	-0.00009849
C	-1.41877907	1.37381417	-0.00004243
C	-2.50296922	0.50771631	0.00016906
C	-2.31026701	-0.88622326	0.00032832
C	-1.02975939	-1.42816272	0.00027740
C	2.21876023	0.71253789	-0.00029684
C	1.15035227	1.54993073	-0.00030447
H	-1.57109706	2.45015525	-0.00016475
H	-3.51296258	0.90804916	0.00021306
H	-3.17165461	-1.54833655	0.00049310
H	-0.88198825	-2.50430519	0.00040046
H	3.26789901	0.97813098	-0.00042800
H	1.23737946	2.63139938	-0.00044990
S	1.76770258	-0.98047157	-0.00004026
Bq	1.01762789	0.31635213	0.99986481
Bq	1.01744195	0.31609816	-1.00013517

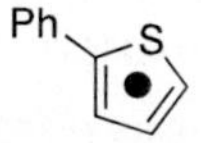

0 1

C	0.76651871	0.41436306	0.63157807
C	1.53714219	1.52099222	0.35527483
C	2.93930284	1.27632616	0.42311061
C	3.23635370	-0.01740036	0.75227275
S	1.79958845	-0.95823459	0.99177924
H	1.10934609	2.47908972	0.08046195
H	3.69310690	2.03112377	0.22603547
H	4.20921484	-0.47519278	0.87128385
C	-0.69801957	0.30267107	0.65691638
C	-1.35082302	-0.91104394	0.37678352
C	-1.48791069	1.42690967	0.96121234
C	-2.74133596	-0.99689908	0.39968430
H	-0.76370567	-1.78875764	0.12003675
C	-2.87799191	1.34151909	0.97131093
H	-1.00361726	2.36622397	1.21160972
C	-3.51273434	0.12902266	0.69286716
H	-3.22312811	-1.94546758	0.17789291
H	-3.46719099	2.22298164	1.21043935
H	-4.59707272	0.06210603	0.70726601
Bq	2.05000036	0.69696076	1.59909581
Bq	2.06156197	0.19745791	-0.33748963

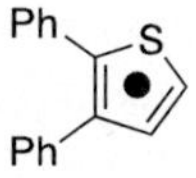

0 1

C	0.73706656	0.97996510	0.00165661
C	-0.63081686	1.20215749	-0.01956266
C	-0.95103481	2.60246205	-0.01576747
C	0.14229522	3.41785602	-0.00200335
S	1.61012763	2.50131490	0.02138406
H	-1.97065487	2.96986902	-0.06117543
H	0.17223716	4.49910022	-0.01795697
C	1.50335542	-0.27844922	0.08052480
C	2.66195552	-0.46623014	-0.69386032
C	1.11366749	-1.30532721	0.95950261
C	3.40431506	-1.64251802	-0.59595503
H	2.97184076	0.31162656	-1.38656617

C	1.85268531	-2.48269486	1.04894478
H	0.23287012	-1.16924400	1.57835349
C	3.00134749	-2.65716314	0.27324440
H	4.29515410	-1.76736938	-1.20584363
H	1.53536790	-3.26301926	1.73569923
H	3.57806027	-3.57506314	0.34832539
C	-1.68802392	0.16361267	-0.09341828
C	-2.83483027	0.26754631	0.71183680
C	-1.60005735	-0.91330805	-0.99173285
C	-3.85667965	-0.67830818	0.62866677
H	-2.91604322	1.08958569	1.41824592
C	-2.62081846	-1.85862798	-1.07353796
H	-0.72995203	-0.99935332	-1.63496620
C	-3.75349249	-1.74615318	-0.26397335
H	-4.73265941	-0.58123927	1.26488302
H	-2.53450149	-2.68172421	-1.77813761
H	-4.54936694	-2.48312518	-0.33039900
Bq	0.16551314	2.13697603	0.99700607
Bq	0.19754195	2.14452619	-1.00272320

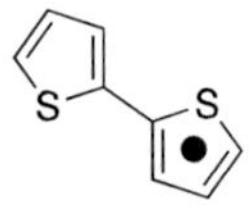

0 1

C	-0.09760225	0.71865042	-0.00004095
C	-1.27480548	1.43579409	-0.00035151
C	-1.08472778	2.84665890	-0.00018393
C	0.23439018	3.20590623	0.00001528
S	1.27480548	1.81591009	0.00026553
H	-2.25144112	0.96303686	-0.00063194
H	-1.89875383	3.56347984	-0.00032920
H	0.66347608	4.19872485	0.00007936
C	0.09760225	-0.71865042	-0.00004095
C	1.27480548	-1.43579409	-0.00035151
S	-1.27480548	-1.81591009	0.00026553
C	1.08472778	-2.84665890	-0.00018393
H	2.25144112	-0.96303686	-0.00063194
C	-0.23439018	-3.20590623	0.00001528
H	1.89875383	-3.56347984	-0.00032920
H	-0.66347608	-4.19872485	0.00007936
Bq	-0.18980993	2.00458220	0.99994086
Bq	-0.18936605	2.00458569	-1.00005909

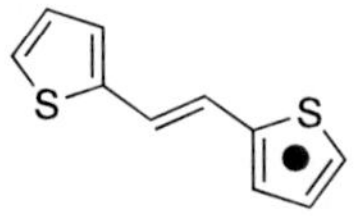

0 1

C	1.89320844	0.36485723	0.00006824
C	2.81450445	1.39361605	0.00051131
C	4.16768934	0.95452498	0.00041514
C	4.28093371	-0.40878819	0.00042714
S	2.72634390	-1.18387637	-0.00004569
H	2.51642170	2.43713404	0.00069505
H	5.02086097	1.62429146	0.00056937
H	5.18088998	-1.00895955	0.00053115
C	0.45550961	0.50047139	-0.00014270
H	0.10832770	1.53367607	-0.00006538
C	-0.45545049	-0.50040359	-0.00022792
H	-0.10822720	-1.53359636	-0.00016875
C	-1.89318652	-0.36482672	-0.00018646
C	-2.81446421	-1.39360686	0.00009213
S	-2.72639294	1.18386638	-0.00033118
C	-4.16767088	-0.95455941	-0.00014646
H	-2.51634819	-2.43711387	0.00026053

C	-4.28096882	0.40874695	-0.00007859
H	-5.02082205	-1.62435162	-0.00012905
H	-5.18094616	1.00888875	-0.00005391
Bq	3.17635599	0.22389635	1.00027519
Bq	3.17671595	0.22423713	-0.99972474

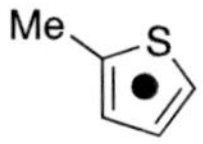

0 1

C	-0.78653977	0.22277281	0.01562729
C	-0.02024184	1.35747410	0.03125821
C	1.38721030	1.10352170	0.00787782
C	1.68258474	-0.22883647	-0.02569882
S	0.23800164	-1.19630286	-0.02907268
H	-0.45226552	2.35311725	0.05898925
H	2.14163072	1.88314112	0.01583049
H	2.65398976	-0.70414739	-0.04858649
C	-2.28198706	0.10505907	0.03186779
H	-2.66394123	-0.40840029	-0.85910062
H	-2.64168112	-0.44820016	0.90815304
H	-2.73074795	1.10311378	0.05998745
Bq	0.51284820	0.22936647	0.99966839
Bq	0.48755780	0.27408525	-0.99967166

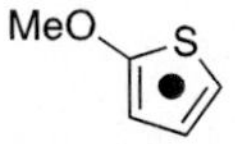

0 1

C	-0.30951680	0.49708801	-0.00651577
C	0.71776974	1.40191208	-0.00378877
C	1.99759550	0.76769164	0.00366488
C	1.93784367	-0.59524873	0.00649825
S	0.28692589	-1.16303781	0.00009928
H	0.55199303	2.47244364	-0.00705389
H	2.93296275	1.31751964	0.00675161
H	2.74597975	-1.31315408	0.01200121
O	-1.62079466	0.82324519	-0.01328979
C	-2.57147871	-0.23415578	-0.01598924
H	-3.55300704	0.24228143	-0.02165158
H	-2.46316722	-0.86188205	-0.91019131
H	-2.47192475	-0.85858051	0.88153632
Bq	0.94724276	0.17172235	0.99971894
Bq	0.93080183	0.17937720	-0.99999483

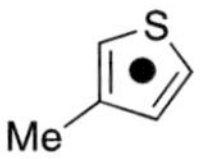

0 1

C	0.08213276	-1.05878612	0.00024230
C	0.98405985	-0.02826167	-0.00039824
C	0.32366567	1.24746030	0.00015224
C	-1.03842709	1.15473021	-0.00006961
S	-1.56272428	-0.50035204	-0.00005215
H	0.85552424	2.19400909	0.00035647
H	-1.76565161	1.95555844	0.00024468
H	0.29129882	-2.12056244	0.00040158
C	2.48018493	-0.19886799	-0.00003841
H	2.93638567	0.26867203	-0.88159923
H	2.93540522	0.26675764	0.88308339
H	2.76092948	-1.25645037	-0.00098221
Bq	-0.24237370	0.16314644	0.99997488
Bq	-0.24214354	0.16276983	-1.00002507

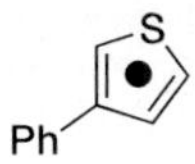

0 1

C	-1.47804901	-0.97217753	-0.31575376
C	-0.65635426	0.07604464	0.02908482
C	-1.42138519	1.23685061	0.40060074
C	-2.77015212	1.04349372	0.33234463
S	-3.15653726	-0.56428952	-0.19480900
H	-0.97161983	2.16539835	0.73468869
H	-3.55999166	1.74157552	0.57482036
H	-1.18671801	-1.95077943	-0.67212826
C	0.82094523	0.02172472	0.01646607
C	1.57940925	1.16593782	-0.28395047
C	1.50605063	-1.17213863	0.30019795
C	2.97251177	1.11692280	-0.30431603
H	1.07196666	2.09572727	-0.52576422
C	2.89831972	-1.22274202	0.27379166
H	0.93991099	-2.06033112	0.56706064
C	3.63901601	-0.07792620	-0.02764407
H	3.53796718	2.01373719	-0.54394175
H	3.40637691	-2.15647109	0.50066321
H	4.72483201	-0.11616388	-0.04338388
Bq	-1.90301026	-0.14483836	1.00139079
Bq	-1.88998090	0.47280714	-0.90080381

References

1. (a) Shirota Y, Kageyama H (2007) Chem Rev 07:953–1010. (b) Wang C, Dong H, Hu W, Liu Y, Zhu D (2012) Chem Rev 112:2208–2267. (c) Sirringhaus H (2014) Adv Mater 26:1319–1335. (d) Facchetti A (2007) Mater Today 10:28–37. (e) Wu W, Liu Y, Zhu D (2010) Chem Soc Rev 39:1489–1502. (f) O'Neill M, Kelly SM (2011) Adv Mater 23:566–584. (g) Murphy AR, Fréchet JMJ (2007) Chem Rev 107:1066–1096. (h) Zhang F, Wu D, Xu Y, Feng X (2011) J Mater Chem 21:17590. (i) Ameri T, Khoram P, Min J, Brabec CJ (2013) Adv Mater 25:4245–4266
2. (a) Kozhevnikov IV (1976) React Kinet Catal Lett 4:451−458. (b) Kozhevnikov IV (1977) React Kinet Catal Lett 6:401−408
3. (a) Masui K, Ikegami H, Mori A (2004) J Am Chem Soc 126:5074–5075. (b) Takahashi M, Masui K, Sekiguchi H, Kobayashi N, Mori A, Funahashi M, Tamaoki N (2006) J Am Chem Soc 128:10930–10933. (c) Kobayashi K, Sugie A, Takahashi M, Masui K, Mori A (2005) Org Lett 7:5083–5085
4. Wang L, Carrow BP (2019) ACS Catal 9:6821–6836
5. Tereniak SJ, Bruns DL, Stahl SS (2020) J Am Chem Soc 142:20318–20323

6. Shang R, Ilies L, Nakamura E (2016) J Am Chem Soc 138:10132–10135
7. Li P, de Bruin B, Reek JNH, Dzik WI (2015) Organometallics 34:5009–5014
8. (a) Ohki Y, Hatanaka T, Tatsumi K (2008) J Am Chem Soc 130:17174–17186. (b) Hatanaka T, Ohki Y, Tatsumi K (2010) Chem Asian J 5:1657–1666
9. Shen K, Fu Y, Li J-N, Liu L, Guo Q-X (2007) Tetrahedron 63:1568–1576
10. Mann G, Baranano D, Hartwig JF, Rheingold AL, Guzei IA (1998) J Am Chem Soc 120:9205–9219
11. Shang R, Ilies L, Nakamura E (2017) Chem Rev 117:9086–9139
12. (a) Hull KL, Sanford MS (2007) J Am Chem Soc 129:11904–11905. (b) Bruns DL, Musaev DG, Stahl SS (2020) J Am Chem Soc 142:19678–19688. (c) Salazar CA, Flesch KN, Haines BE, Zhou PS, Musaev DG, Stahl SS (2020) Science 370:1454–1460
13. Kuo Y-N, Chen F, Ainsworth C, Bloomfield JJ (1971) J Chem Soc D 136–137
14. Shang R, Ilies L, Nakamura E (2015) J Am Chem Soc 137:7660–7663
15. (a) Grzybowski M, Skonieczny K, Butenschön H, Gryko DT (2013) Angew Chem Int Ed 52:9900–9930. (b)Watson MD, Fechtenkötter A, Müllen K (201) Chem Rev 101:1267–1300
16. Roncali J, Blanchard P, Frère P (2005) J Mater Chem 15:1589–1610
17. Wu Y, Wang Z, Liang M, Cheng H, Li M, Liu L, Wang B, Wu J, Prasad Ghimire R, Wang X, Sun Z, Xue S, Qiao Q (2018) ACS Appl Mater Interfaces 10:17883–17895
18. (a) Schleyer PVR, Maerker C, Dransfeld A, Jiao H, van Eikema Hommes NJR (1996) J Am Chem Soc 118:6317–6318. (b) Schleyer PVR, Manoharan M, Wang Z-X, Kiran B, Jiao H, Puchta R, van Eikema Hommes NJR (2001) Org Lett 3:2465–2468
19. (a) Xu J, Bercher OP, Watson MP (2021) J Am Chem Soc 143:8608–8613. (b) Tobisu M, Shimasaki T, Chatani N (2008) Angew Chem Int Ed 47:4866–4869. (c) Yu D-G, Shi Z-J (2011) Angew Chem Int Ed 50:7097–7100. (d) Guo L, Liu X, Baumann C, Rueping M (2016) Angew Chem Int Ed 55:15415–15419. (e) Koch E, Takise R, Studer A, Yamaguchi J, Itami K (2015) Chem Commun 51:855–857. (f) Wang T-H, Ambre R, Wang, Q, Lee W-C, Wang P-C, Liu Y, Zhao L, Ong T-G (2018) ACS Catal 8:11368–11376
20. Simmons EM, Hartwig JF (2012) Angew Chem Int Ed 51:3066–3072
21. Still WC, Kahn M, Mitra A (1978) J Org Chem 43:2923–2925
22. Pangborn AB, Giardello MA, Grubbs RH, Rosen RK, Timmers FJ (1996) Organometallics 15:1518–1520
23. Doba T, Ilies L, Sato W, Shang R, Nakamura E (2021) Nat Catal 4:631–638
24. Truong T, Alvarado J, Tran LD, Daugulis O (2010) Org Lett 12:1200–1203
25. Schroth W, Hintzsche E, Jordan H, Jende T, Spitzner R, Thondorf I (1997) Tetrahedron 53:7509–7528
26. Yassar A, Garnier F, Deloffre F, Horowitz G, Ricard L (1994) Adv Mater 6:660–663
27. Ball M, Zhong Y, Fowler B, Zhang B, Li P, Etkin G, Paley DW, Decatur J, Dalsania AK, Li H, Xiao S, Ng F, Steigerwald ML, Nuckolls C (2016) J Am Chem Soc 138:12861–12867
28. Holzer B, Bintinger J, Lumpi D, Choi C, Kim Y, Stöger B, Hametner C, Marchetti-Deschmann M, Plasser F, Horkel E, Kymissis I, Fröhlich J (2017) ChemPhysChem 18:549–563
29. Song J, Wei F, Sun W, Cao X, Liu C, Xie L, Huang W (2014) Org. Chem. Front. 1:817–820
30. Skakovskii ED, Stankevich AI, Tychinskaya LY, Shirokii OV, Choban YP, Murashko VL, Rykov S V (2004) Zh Obshch Khim 74:1719–1725
31. Gaussian 09, Revision B.01, Frisch MJ, Trucks GW, Schlegel HB, Scuseria GE, Robb MA, Cheeseman JR, Scalmani G, Barone V, Mennucci B, Petersson GA, Nakatsuji H, Caricato M, Li X, Hratchian HP, Izmaylov AF, Bloino J, Zheng G, Sonnenberg JL, Hada M, Ehara M, Toyota K, Fukuda R, Hasegawa J, Ishida M, Nakajima T, Honda Y, Kitao O, Nakai H, Vreven T, Montgomery Jr JA, Peralta JE, Ogliaro F, Bearpark M, Heyd JJ, Brothers E, Kudin KN, Staroverov VN, Keith T, Kobayashi R, Normand J, Raghavachari K, Rendell A, Burant JC, Iyengar SS, Tomasi J, Cossi M, Rega N, Millam JM, Klene M, Knox JE, Cross JB, Bakken V, Adamo C, Jaramillo J, Gomperts R, Stratmann RE, Yazyev O, Austin AJ, Cammi R, Pomelli C, Ochterski JW, Martin RL, Morokuma K, Zakrzewski VG, Voth GA, Salvador P, Dannenberg JJ, Dapprich S, Daniels AD, Farkas O, Foresman JB, Ortiz JV, Cioslowski J, Fox DJ (2010) Gaussian, Inc., Wallingford CT

Chapter 3
Development of a Synthetic Method for Tridentate Phosphine Ligands

3.1 Introduction

Tridentate phosphine compounds have been reported to be an effective ligand for various kinds of transition-metal-catalyzed reactions such as iron-catalyzed C–H activation [1], iron-catalyzed nitrogen reduction [2], cobalt-catalyzed addition to alkynes [3], cobalt-catalyzed hydrogenation of carboxylic acids [4], ruthenium-catalyzed hydrogenation [5], and molybdenum-catalyzed nitrogen reduction [6]. In addition to transition-metal catalysis, their copper complexes have been reported to be useful for OLEDs [7]. Currently, the synthetic methods for conjugated tridentate phosphine ligands rely on the use of phosphine chlorides that are toxic, difficult to handle, and limited in access (Scheme 3.1).

To accelerate the research in which tridentate phosphine ligands are used, a modular and practical method to synthesize tridentate phosphine ligands is desirable. In this work, I developed a simple and scalable method to synthesize tridentate phosphine ligands by sequential addition of organolithium reagents to $P(OPh)_3$ (Scheme 3.2). The use of phosphine chlorides was eliminated, and all of the aryl groups on three phosphines were introduced in a controlled manner by lithium-halogen exchange or lithiation of the corresponding starting materials.

3.2 Selective Formation of Phenoxydiarylphosphine

To achieve clean formation of the desired tridentate phosphine ligand by sequential addition of organolithium reagents to an electrophilic phosphine reagent (PX_3), the first reaction of an organolithium reagent with an electrophilic phosphine reagent needs to selectively generate a tridentate phosphine precursor (**II**) out of bidentate (**I**) and tetradentate (**III**) ones (Scheme 3.3).

T. Doba, *Iron-Catalyzed C–H/C–H Coupling for Synthesis of Functional Small Molecules and Polymers*, Springer Theses,
https://doi.org/10.1007/978-981-99-4121-6_3

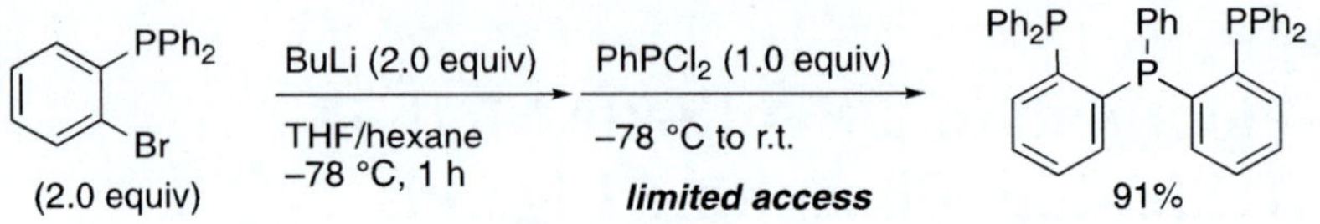

Scheme 3.1 Synthesis of a conjugated tridentate phosphine ligand

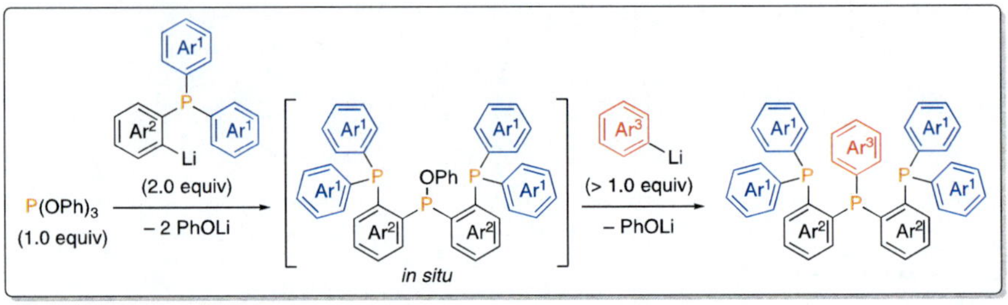

Scheme 3.2 Synthesis of tridentate phosphine ligands by sequential addition of organolithium reagents to $P(OPh)_3$

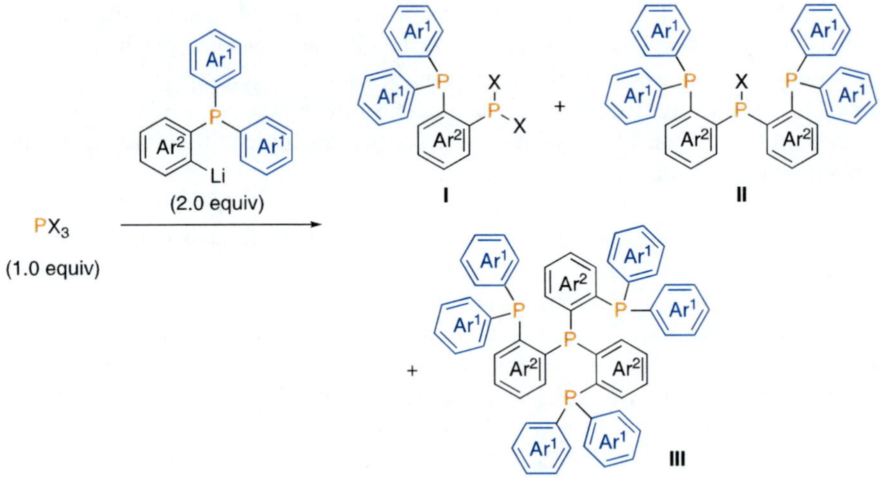

Scheme 3.3 Selectivity issue in the addition of the first organolithium reagent

Knowing this potential difficulty in this strategy, I thought that $P(OPh)_3$ can be used as the electrophilic phosphine to selectively form **II** based on the report by Straub and coworkers that $P(OPh)_3$ can undergo selective stepwise nucleophilic substitution in one-pot (Scheme 3.4)[8].

Following their procedure, 2.0 equiv of (2-bromophenyl)diphenylphosphane (**1**) was treated with 2.0 equiv of BuLi for lithium-halogen exchange, and the generated (2-(diphenylphosphaneyl)phenyl)lithium (**2**) was reacted with 1.0 equiv of $P(OPh)_3$

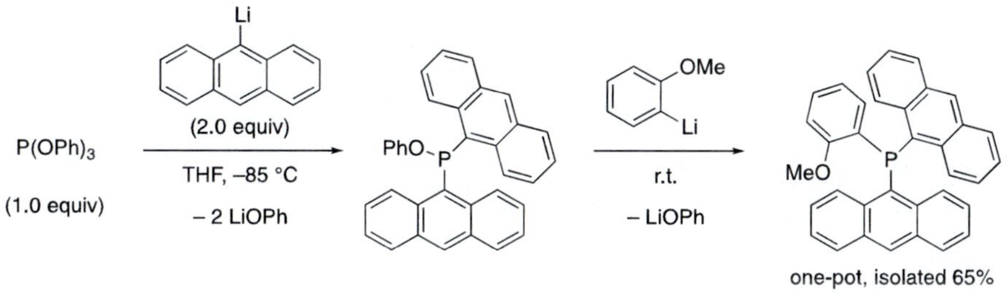

Scheme 3.4 One-pot synthesis of sterically shielded phosphorus ligands by selective stepwise nucleophilic substitution at $P(OPh)_3$

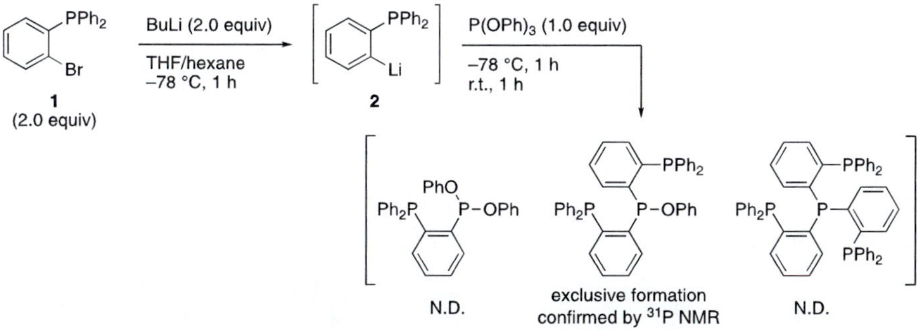

Scheme 3.5 Synthesis of a tridentate phosphine ligand precursor by treatment of $P(OPh)_3$ with (2-(diphenylphosphaneyl)phenyl)lithium

at −78 °C. After gradually warming to room temperature, the crude reaction mixture was analyzed by ^{31}P NMR. The NMR spectrum supported the exclusive formation of the tridentate phosphine precursor and absence of the bidentate and tetradentate ones (Scheme 3.5, Fig. 3.1). This indicates that subsequent reaction of the crude reaction mixture with another organolithium reagent will cleanly afford the tridentate phosphine ligand.

3.3 Synthesis of Tridentate Phosphine Ligands

With the tridentate phosphine ligand precursor in hand, a subsequent reaction of the precursor with various kinds of organolithium reagents was attempted to synthesize tridentate phosphine ligands as a final product. As shown in Table 3.1, the tridentate phosphine ligand precursor was treated with 1.5 equiv of aryl lithium reagents generated by lithium-halogen exchange of the corresponding aryl bromides with BuLi. The pure tridentate phosphine ligands were obtained in high yields by recrystallization from dichloromethane/alcohol. This method is applicable to various kinds of aryl groups ranging from electron rich to deficient ones (**3–6**). This method is also

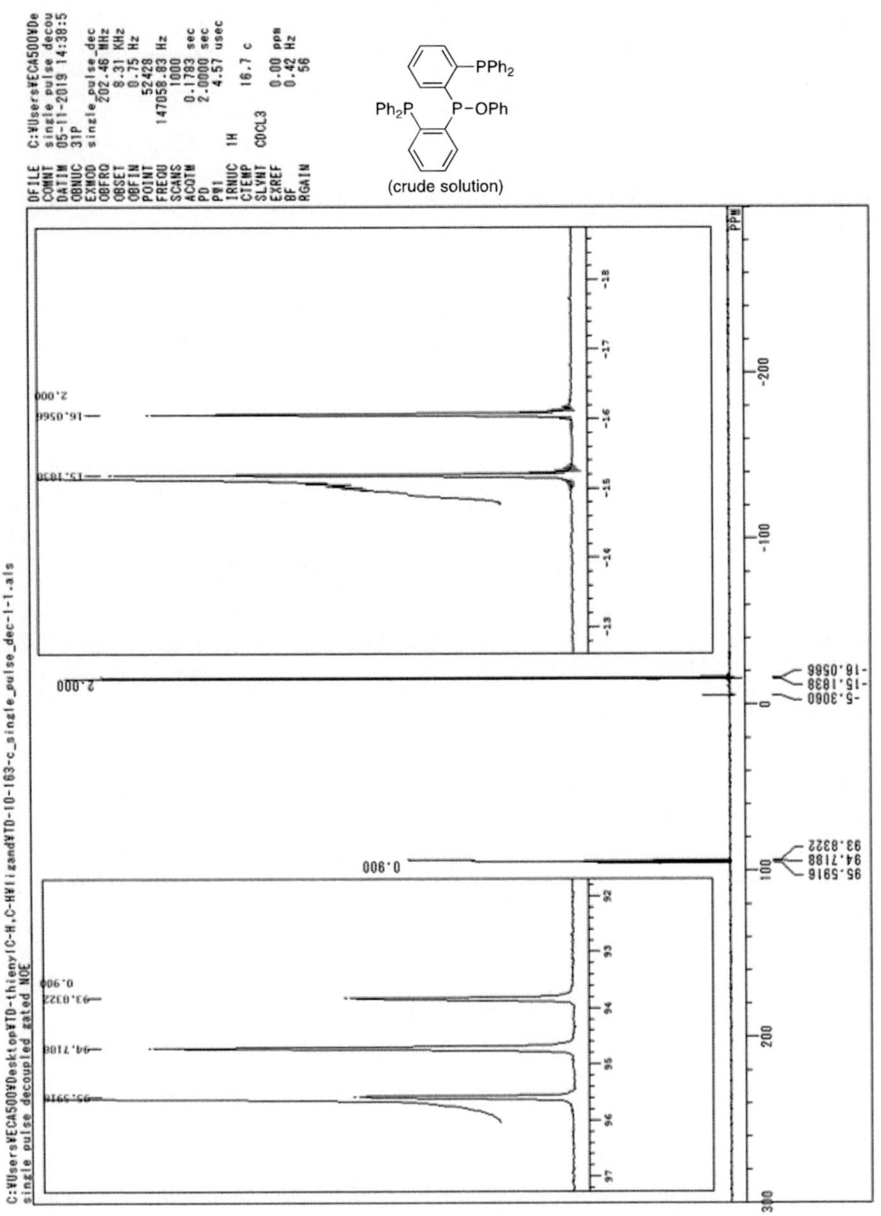

Fig. 3.1. ^{31}P NMR spectrum of the crude reaction mixture after the first organolithium addition to $P(OPh)_3$

useful for the synthesis of a deuterium-labeled tridentate phosphine ligand which is difficult to be synthesized by any other synthetic methods (**7**). Notably, an *ortho*-substituted organolithium reagent was reactive (**8**), opening up the possibility of using this method even for the synthesis of tetradentate ligands.

Furthermore, the synthesis of heteroaryl-TP was attempted by reacting the tridentate phosphine ligand precursor with 1.5 equiv of heteroaryl lithium reagents generated by lithiation of the corresponding heteroarenes with BuLi. Also in this case, the pure products were obtained in high yields by simple recrystallization. TMEDA was added as an additive for the lithiation of 1-methylindole to afford indolyl-TP (**9**). This method was widely effective for the synthesis of benzofuryl-TP (**10**). benzothienyl-TP (**11**), benzimidazolyl-TP (**12**), and benzothiazolyl-TP (**13**). However, the synthesis of benzoxazolyl-TP (**14**) failed possibly because of the ring opening of benzoxazolyl lithium reagent [9]. Gram-scale syntheses were

Table 3.1 Synthesis of aryl-TPs

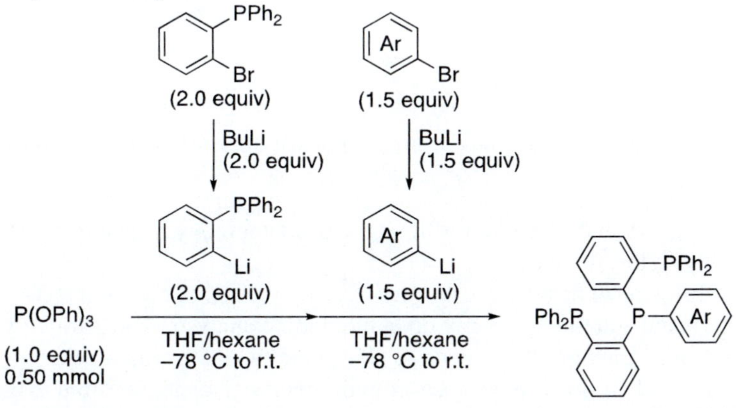

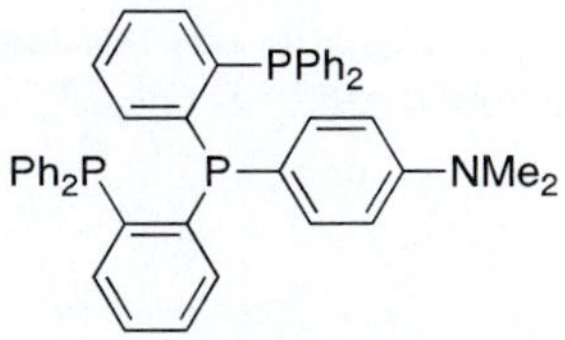

3 (Me_2N-TP)
88%

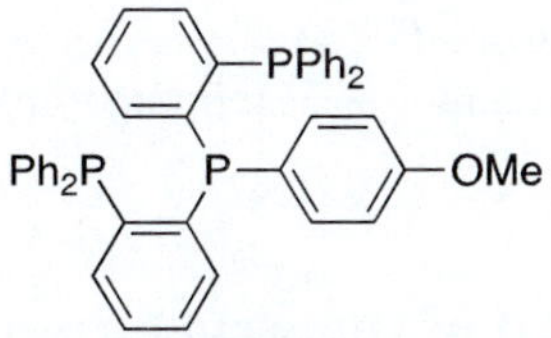

4 (MeO-TP)
85%

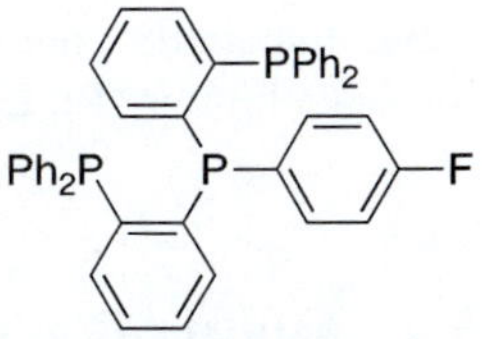

5 (F-TP)
85%

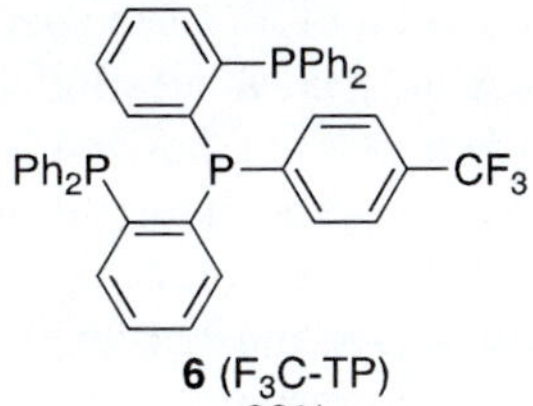

6 (F_3C-TP)
83%

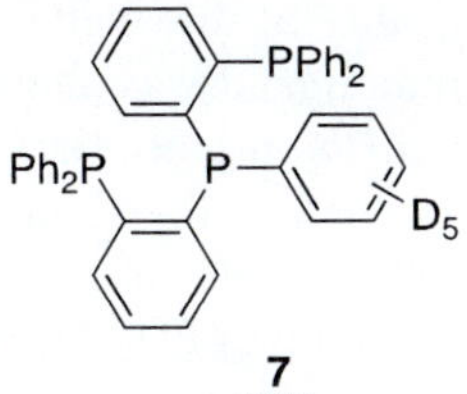

7
87%

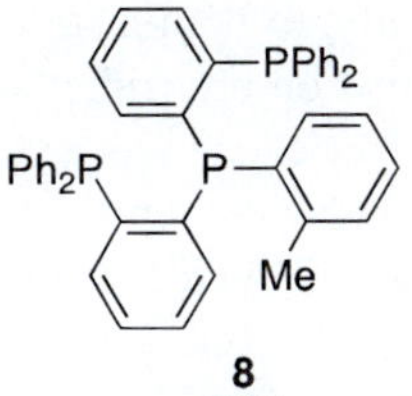

8
85%

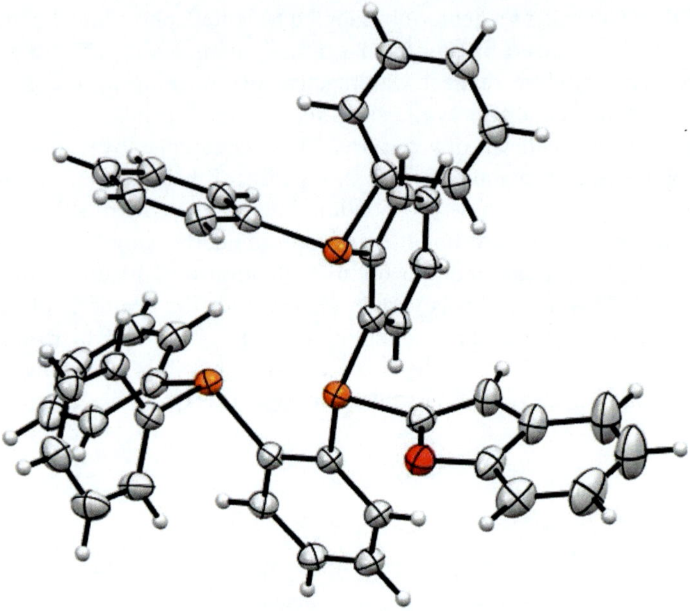

Fig. 3.2 ORTEP drawing of benzofuryl-TP (**10**). Thermal ellipsoids are shown at 30% probability

achieved with **10** and **11** without silica gel chromatography, which demonstrates the practicality of this method.

Slow diffusion of hexane into an ethyl acetate solution of benzofuryl-TP (**10**) gave a single crystal suitable for X-ray crystallographic analysis. The X-ray crystal data confirmed the structure of a tridentate phosphine ligand having a benzofuryl group as a central aryl group. The distance from the central phosphorus atom to the closest hydrogen atom of the central aryl group was determined to be 3.14 Å, which is longer than in the case of Me_2N-TP (**3**) (2.80 Å). This feature may help to create less sterically hindered catalytic center and facilitate transition-metal catalysis (Fig. 3.2).

3.4 Modulation of All of the Aryl Groups

Finally, to demonstrate the flexibility of this method, modulation of all of the aryl groups on the tridentate phosphine ligand was attempted. A fully *meta*-substituted TP was chosen as a target. First, *ortho*-bromo triarylphosphine was synthesized by following the reported two-step procedure: Palladium-catalyzed C–P bond formation between aryl triflate (**15**) and diarylphosphine oxide (**16**), and subsequent reduction of triarylphosphine oxide to triarylphosphine (**17**) [10]. Aryl triflate (**15**) and diarylphosphine oxide (**16**) starting materials were synthesized from the corresponding Grignard reagent and phenol, respectively. Thus, this two-step procedure allows us to

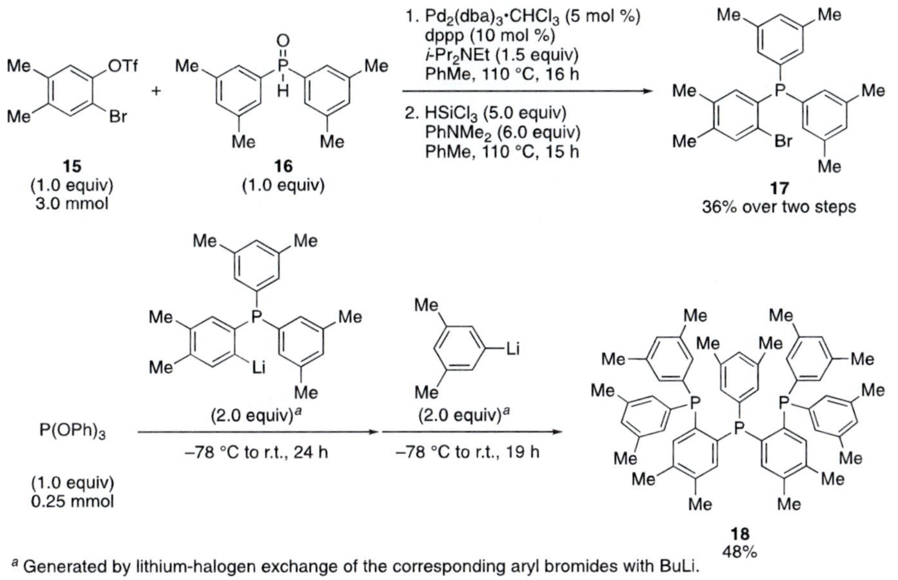

[a] Generated by lithium-halogen exchange of the corresponding aryl bromides with BuLi.

Scheme 3.6 Synthesis of a fully meta-substituted TP

introduce various kinds of substituents to the *ortho*-bromo triarylphosphines starting from readily available aryl bromides and phenols. Finally, sequential addition of two organolithium reagents generated from the *ortho*-bromo triarylphosphine and an aryl bromide to $P(OPh)_3$ afforded the fully *meta*-substituted TP as a pure product (Scheme 3.6).

3.5 Conclusion

In conclusion, various kinds of conjugated tridentate phosphine ligands were synthesized by sequential addition of two different organolithium reagents to $P(OPh)_3$. The use of $P(OPh)_3$ as a electrophilic phosphine source was the key to selectively synthesize tridentate phosphine ligand out of bidentate or tetradentate ones. This method was applicable to the synthesis of aryl-TPs and heteroaryl-TPs, and the pure products were obtained on a gram scale by simple recrystallization from common organic solvents. This method helps to accelerate the investigation of catalytic reactions and functional complexes using tridentate phosphine ligands. The use of heteroaryl-TPs for iron-catalyzed regioselective thienyl C–H/C–H polycondensation will be discussed in the next chapter.

3.6 Experimental

Materials and methods

All air or moisture-sensitive reactions were performed in a dry reaction vessel under argon atmosphere. Air or moisture-sensitive liquids and solutions were transferred with syringe or Teflon cannula. The water content of solvents was confirmed to be less than 30 ppm by Karl Fischer titration performed with MKC-210 (Kyoto Electronics Manufacturing Co., Ltd.). Analytical thin-layer chromatography (TLC) was performed with a glass plate coated with 0.25 mm 230–400 mesh silica gel containing a fluorescent indicator. Organic solutions were evacuated with a diaphragm pump through a rotary evaporator. Flash column chromatography was performed as described by Still et al. [11] Preparative recycling gel permeation chromatography (GPC) was performed with LC-92XX II NEXT instrument (Japan Analytical Industry Co., Ltd.) equipped with JAIGEL-2 h polystyrene columns using chloroform as an eluent at the flow rate of 7.5 mL/min.

Gas chromatography (GC) was performed with GC-2014 instrument (Shimadzu Co.) equipped with an ULBON HR-1 (0.25 mm I.D. × 25 mL, 0.25 μm, Shinwa Chemical Industries, Ltd.) capillary column. Mass spectra (GC–MS) were taken with Parvum 2 instrument (Shimadzu Co.). High-resolution mass spectra (HRMS) were taken with LCMS-IT-TOF (Shimadzu Co.) using reserpine (MW 608.2734) as an internal standard. Melting points of solid compounds were measured on a Mel-Temp capillary melting-point apparatus and were uncorrected. Nuclear magnetic resonance (NMR) spectra were taken with ECZ-500 (JEOL, Ltd.) at room temperature unless otherwise noted and reported in parts per million (ppm). ^{1}H NMR spectra were internally referenced to tetramethylsilane (0.00 ppm), $CHCl_3$ (7.26 ppm), $CHDCl_2$ (5.32 ppm), C_2HDCl_4 (5.97 ppm), or $(CHD_2)(CD_3)SO$ (2.50 ppm). ^{13}C NMR spectra were internally referenced to tetramethylsilane (0.0 ppm), $CDCl_3$ (77.0 ppm), CD_2Cl_2 (53.8 ppm), $C_2D_2Cl_4$ (73.8 ppm), or $(CD_3)_2SO$ (39.5 ppm). ^{19}F NMR spectra were internally referenced to C_6F_6 (–164.9 ppm). ^{31}P NMR spectra were internally referenced to $(CH_3O)_3PO$ (2.1 ppm). ICP analysis was performed on Shimadzu ICPS-7510 equipment.

Unless otherwise noted, reagents were purchased from Tokyo Chemical Industry Co., Ltd., Sigma-Aldrich Co., LCC, FUJIFILM Wako Pure Chemical Co., and other commercial suppliers and were used as received. Anhydrous tetrahydrofuran and diethyl ether were purchased from KANTO Chemical Co., Inc. and purified prior to use by a solvent purification system (GlassContour) equipped with columns of activated alumina and supported copper catalyst [12]. (2-bromophenyl)diphenylphosphane was purchased from Tokyo Chemical Industry Co., Ltd. and used as received. A tetrahydrofuran solution of butyllithium was purchased from Tokyo Chemical Industry Co., Ltd and titrated prior to use.

Preparation of the crude solution of ((phenoxyphosphanediyl)bis(2,1-phenylene))bis(diphenylphosphane) (Scheme 3.5)

A hexane solution of BuLi (1.60 mol/L, 3.75 mL, 6.00 mmol) was added dropwise to a solution of (2-bromophenyl)diphenylphosphane (2.05 g, 6.00 mmol) in THF (12 mL) at −78 °C. After stirring for 1 h, $P(OPh)_3$ (0.78 mL, 3.0 mmol) dissolved in THF (6.0 mL) was added dropwise and the reaction mixture was stirred at −78 °C for 1 h and then at rt for 1 h. ^{31}P NMR analysis of the crude mixture showed a clean formation of ((phenoxyphosphanediyl)bis(2,1-phenylene))bis(diphenylphosphane). The obtained solution was used as a 0.133 mol/L (= 3.00 mmol/22.5 mL) solution of ((phenoxyphosphanediyl)bis(2,1-phenylene))bis(diphenylphosphane) in the next step. The solution can be stored at rt under argon over one week.

A general synthetic procedure for aryl-TPs (Table 3.1, compound 3–8)

A hexane solution of BuLi (1.53 mol/L, 0.49 mL, 0.75 mmol) was added dropwise to a solution of an aryl bromide (0.75 mmol) in THF (1.5 mL) at −78 °C. After stirring for 1 h, the crude solution of ((phenoxyphosphanediyl)bis(2,1-phenylene))bis(diphenylphosphane) (0.132 mol/L, 3.8 mL, 0.50 mmol) was added dropwise, and the reaction mixture was stirred at rt for 18 h. The reaction mixture was quenched by addition of water, and the aqueous layer was extracted with dichloromethane. The combined organic layers were dried over Na_2SO_4 and passed through a pad of Florisil. The solvent was removed under reduced pressure, and the pure product was obtained by recrystallization.

4-(Bis(2-(diphenylphosphaneyl)phenyl)phosphaneyl)-*N*,*N*-dimethylaniline (3):

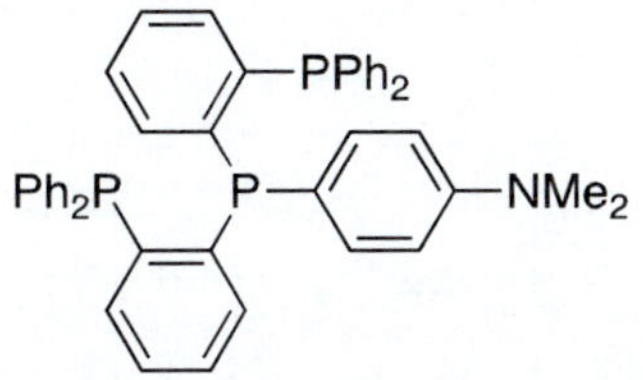

The title compound was obtained as white solid in 88% yield on a 0.5 mmol scale using 4-bromo-*N*,*N*-dimethylaniline as the aryl bromide. Dichloromethane/methanol was used for recrystallization. The compound data was in good agreement with the literature [1].

^{1}H NMR (500 MHz, $CDCl_3$): δ 7.26–6.97 (m, 28H), 6.90–6.87 (m, 2H), 6.50–6.49 (m, 2H), 2.92 (s, 6H).

^{31}P NMR (202 MHz, $CDCl_3$): δ −15.7 (AB_2m, J = 151 Hz, 2P), −20.4 (AB_2m, J = 151 Hz, 1P).

(((4-Methoxyphenyl)phosphanediyl)bis(2,1-phenylene))bis(diphenylphosphane) (4):

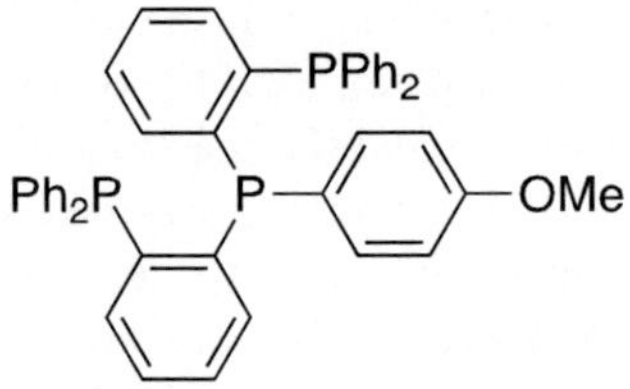

The title compound was obtained as white solid in 85% yield on a 0.5 mmol scale using 1-bromo-4-methoxybenzene as the aryl bromide. Dichloromethane/methanol was used for recrystallization.

Melting point: 231–232 °C (methanol).

^{1}H NMR (500 MHz, $CDCl_3$): δ 7.26–7.02 (m, 28H), 6.86–6.82 (m, 2H), 6.68–6.67 (m, 2H), 3.76 (s, 3H).

^{31}P NMR (202 MHz, $CDCl_3$): δ −15.5 (AB_2m, J = 153 Hz, 2P), −20.6 (AB_2m, J = 153 Hz, 1P).

HRMS (APCI +): *m/z* calcd for $C_{43}H_{35}OP_3$ [M + H$^+$] 661.1974; found: 661.1949.

(((4-Fluorophenyl)phosphanediyl)bis(2,1-phenylene))bis(diphenylphosphane) (5):

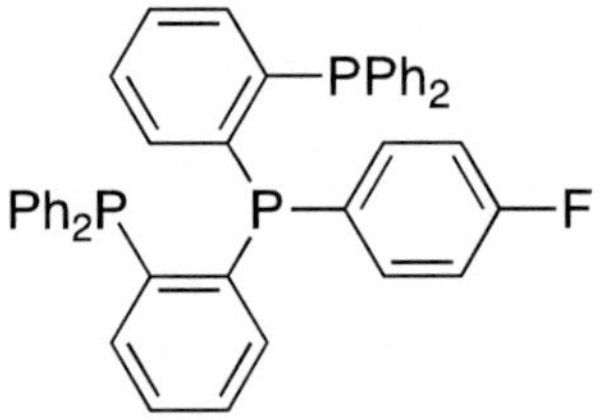

The title compound was obtained as white solid in 85% yield on a 0.5 mmol scale using 1-bromo-4-fluorobenzene as the aryl bromide. Dichloromethane/methanol was used for recrystallization.

Melting point: 215–217 °C (methanol).

^{1}H NMR (500 MHz, $CDCl_3$): δ 7.26–7.03 (m, 28H), 6.83–6.79 (m, 4H).

^{31}P NMR (202 MHz, $CDCl_3$): δ −15.4 (AB_2m, J = 155 Hz, 2P), −20.9 (AB_2m, J = 155 Hz, 1P).

^{19}F NMR (471 MHz, $CDCl_3$): δ −116.7.

HRMS (APCI +): *m/z* calcd for $C_{42}H_{32}FP_3$ [M + H$^+$] 649.1774; found: 649.1778.

(((4-(Trifluoromethyl)phenyl)phosphanediyl)bis(2,1-phenylene))bis(diphenylphosphane) (6):

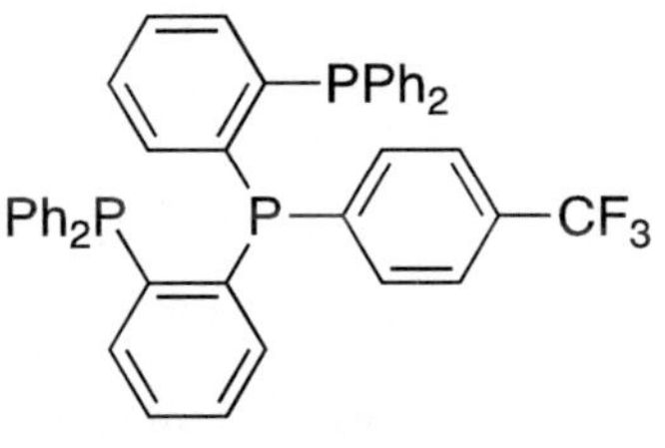

The title compound was obtained as pale orange solid in 83% yield on a 0.5 mmol scale using 1-bromo-4-(trifluoromethyl)benzene as the aryl bromide. Dichloromethane/ethanol was used for recrystallization.

Melting point: 240–241 °C (ethanol).

^{1}H NMR (500 MHz, $CDCl_3$): δ 7.30–7.29 (m, 2H), 7.26–7.07 (m, 28H), 6.81–6.80 (m, 2H).

^{31}P NMR (202 MHz, $CDCl_3$): δ −15.2 (AB_2m, J = 155 Hz, 2P), −20.2 (AB_2m, J = 155 Hz, 1P).

^{19}F NMR (471 MHz, $CDCl_3$): δ −65.9.

HRMS (APCI +): m/z calcd for $C_{43}H_{32}F_3P_3$ [M + H^+] 699.1742; found: 699.1755.

(((Phenyl-*d*5)phosphanediyl)bis(2,1-phenylene))bis(diphenylphosphane) (7):

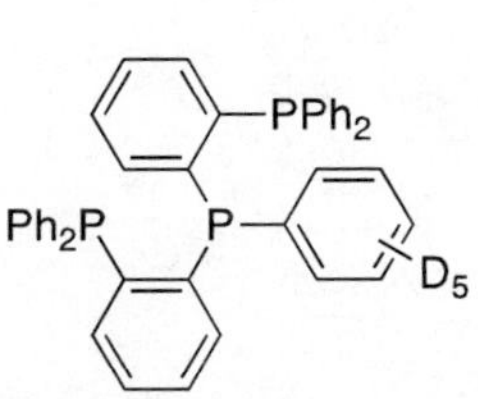

The title compound was obtained as white solid in 87% yield on a 0.5 mmol scale using 1-bromobenzene-2,3,4,5,6-*d*5 as the aryl bromide. Dichloromethane/methanol was used for recrystallization.

Melting point: 217–218 °C (methanol).

^{1}H NMR (500 MHz, $CDCl_3$): δ 7.26–7.06 (m, 26H), 6.84–6.83 (m, 2H).

^{31}P NMR (202 MHz, $CDCl_3$): δ −15.3 (AB_2m, J = 155 Hz, 2P), −19.2 (AB_2m, J = 155 Hz, 1P).

HRMS (APCI +): m/z calcd for $C_{42}H_{28}D_5P_3$ [M + H^+] 636.2182; found: 636.2183.

((*o*-Tolylphosphanediyl)bis(2,1-phenylene))bis(diphenylphosphane) (8):

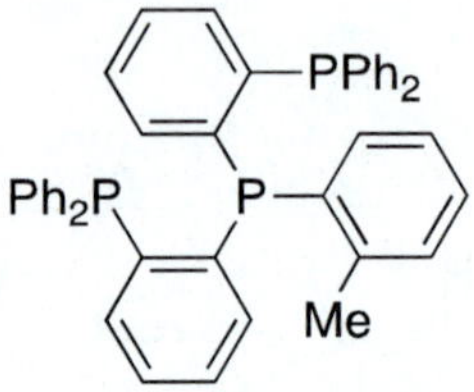

The title compound was obtained as white solid in 85% yield on a 0.5 mmol scale using 1-bromo-2-methylbenzene as the aryl bromide. Dichloromethane/methanol was used for recrystallization.

Melting point: 229–232 °C (methanol).

^{1}H NMR (500 MHz, $CDCl_3$): δ 7.26–7.04 (m, 28H), 6.91–6.85 (m, 3H), 6.66–6.64 (m, 1H), 2.09 (s, 3H).

^{31}P NMR (202 MHz, $CDCl_3$): δ −14.3 (d, $J = 156$ Hz, 2P), −25.6 (t, $J = 156$ Hz, 1P).

HRMS (APCI +): *m/z* calcd for $C_{43}H_{35}P_3$ [M + H$^+$] 645.2024; found: 645.2015.

2-(Bis(2-(diphenylphosphaneyl)phenyl)phosphaneyl)-1-methyl-1*H*-indole (Table 3.2, compound 9):

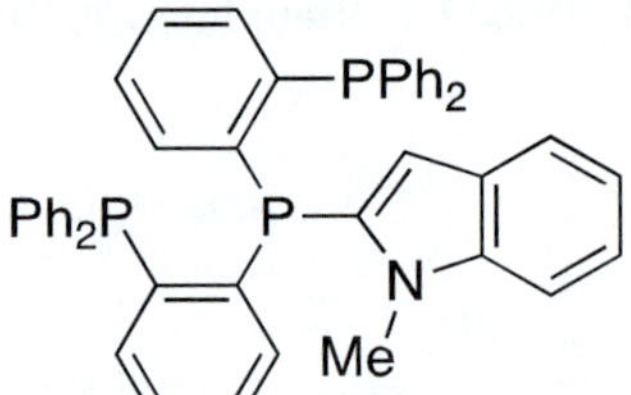

A hexane solution of BuLi (1.53 mol/L, 0.49 mL, 0.75 mmol) was added dropwise to a mixture of 1-methyl-1*H*-indole (94 μL, 0.75 mmol) and TMEDA (112 μL, 0.75 mmol) in diethyl ether (1.5 mL) at rt. After stirring for 4 h, the crude solution of ((phenoxyphosphanediyl)bis(2,1-phenylene))bis(diphenylphosphane) (0.132 mol/L, 3.8 mL, 0.50 mmol) was added dropwise at 0 °C and the reaction mixture was stirred at rt for 10 h. The reaction mixture was quenched by addition of water (10 mL) and the aqueous layer was extracted with dichloromethane (50 mL). The combined organic layers were washed with brine, dried over Na_2SO_4, and passed through a pad of Florisil. The solvent was removed under reduced pressure, and the crude product was carefully recrystallized from dichloromethane/methanol to afford the product as white solid (0.28 g, 82%).

Melting point: 239–240 °C (methanol).

Table 3.2 Synthesis of heteroaryl-TPs

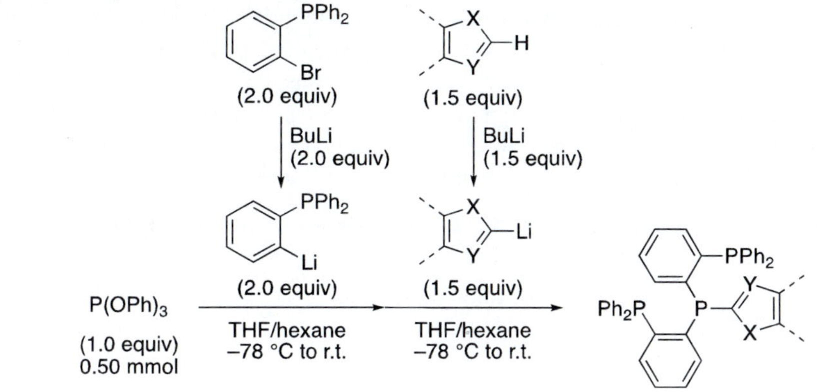

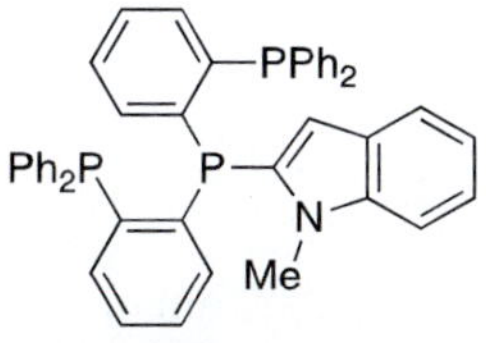

9 (Indolyl-TP)
82%[a]

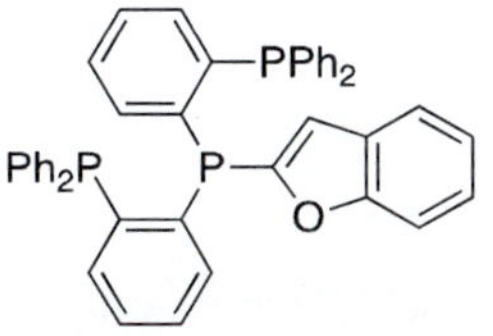

10 (Benzofuryl-TP)
90% (1.80 g)

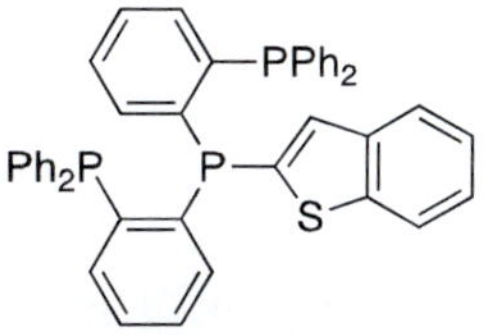

11 (Benzothienyl-TP)
85% (1.16 g)

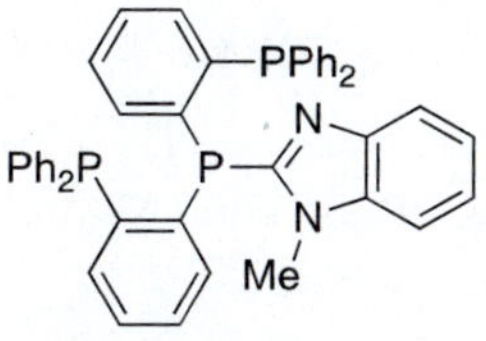

12 (Benzimidazolyl-TP)
62%

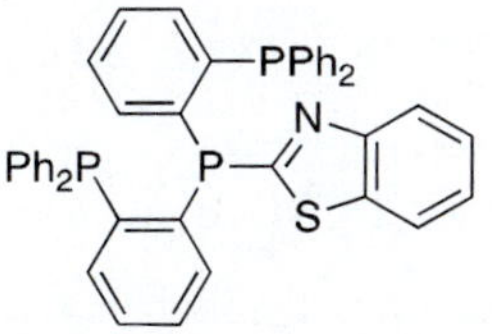

13 (Benzothiazolyl-TP)
69%

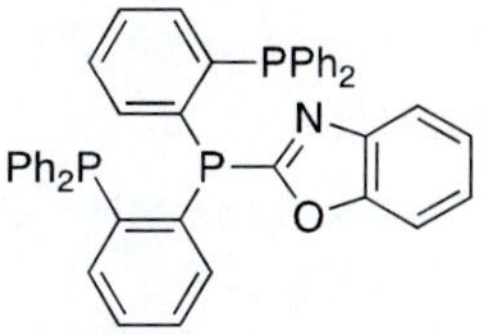

14 (Benzoxazolyl-TP)
0%

[a] BuLi (1.5 equiv)/TMEDA (1.5 equiv) in Et_2O at r.t. for the deprotonation of 1-methylindole.

^{1}H NMR (500 MHz, $CDCl_3$): δ 7.39 (d, J = 7.7 Hz, 1H), 7.26–6.96 (m, 31H), 5.91 (s, 1H), 3.45 (s, 3H).

^{31}P NMR (202 MHz, $CDCl_3$): δ −13.7 (d, J = 154 Hz, 2P), −42.8 (t, J = 154 Hz, 1P).

HRMS (APCI +): *m/z* calcd for $C_{45}H_{36}NP_3$ [M + H$^+$] 684.2133; found: 684.2121.

A general synthetic procedure for heteroaryl-TPs (Table 3.2, compound 10–13)

A hexane solution of BuLi (1.53 mol/L, 1.96 mL, 3.00 mmol) was added dropwise to a solution of a heteroarene (3.00 mmol) in THF (6.0 mL) at −78 °C.

After stirring for 1 h, the crude solution of ((phenoxyphosphanediyl)bis(2,1-phenylene))bis(diphenylphosphane) (0.132 mol/L, 15 mL, 2.0 mmol) was added dropwise, and the reaction mixture was stirred at rt for 12 h. The reaction mixture was quenched by addition of water, and the aqueous layer was extracted with dichloromethane. The combined organic layers were washed with brine, dried over Na_2SO_4, and passed through a pad of Florisil. The solvent was removed under reduced pressure, and the pure product was obtained by recrystallization.

((Benzofuran-2-ylphosphanediyl)bis(2,1-phenylene))bis(diphenylphosphane) (10):

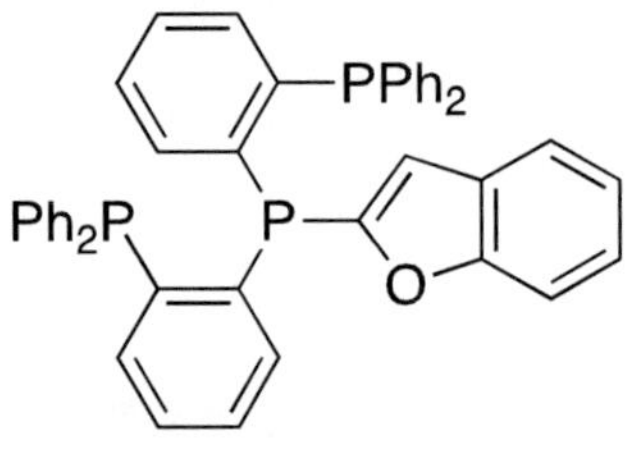

The title compound was obtained as white solid in 90% yield on a 3.0 mmol scale using 2,3-benzofuran as the heteroarene. Dichloromethane/methanol was used for recrystallization.

Melting point: 246–248 °C (methanol).

^{1}H NMR (500 MHz, $CDCl_3$): δ 7.36–7.33 (m, 2H), 7.23–7.06 (m, 30H), 6.44–6.43 (m, 1H).

^{31}P NMR (202 MHz, $CDCl_3$): δ −14.7 (AB_2m, J = 154 Hz, 2P), −37.6 (AB_2m, J = 154 Hz, 1P).

HRMS (APCI +): *m/z* calcd for $C_{44}H_{33}OP_3$ [M + H^+] 671.1817; found: 671.1816.

((Benzo[*b*]thiophen-2-ylphosphanediyl)bis(2,1-phenylene))bis(diphenylphosphane) (11):

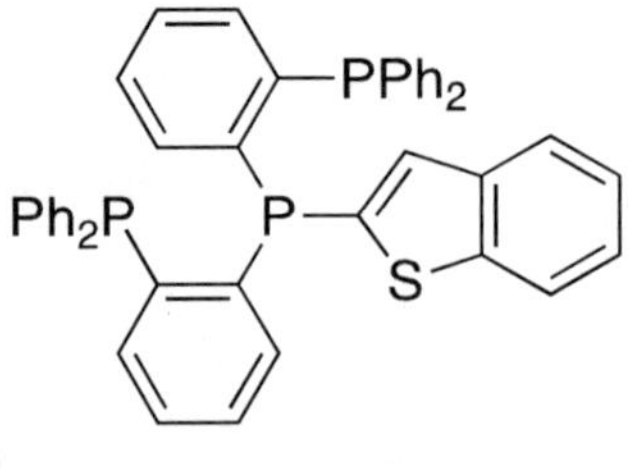

The title compound was obtained as white solid in 85% yield on a 2.0 mmol scale using benzo[*b*]thiophene as the heteroarene. Dichloromethane/methanol was used for recrystallization.

Melting point: 259–260 °C (methanol).

^{1}H NMR (500 MHz, CD_2Cl_2): δ 7.65–7.63 (m, 1H), 7.55–7.53 (m, 1H), 7.24–6.97 (m, 31H).

^{31}P NMR (202 MHz, CD_2Cl_2): δ −14.7 (AB_2m, J = 154 Hz, 2P), −30.1 (AB_2m, J = 154 Hz, 1P).

HRMS (APCI +): *m/z* calcd for $C_{44}H_{33}P_3S$ [M + H$^+$] 687.1589; found: 687.1567.

2-(Bis(2-(diphenylphosphaneyl)phenyl)phosphaneyl)-1-methyl-1*H*-benzo[*d*]imidazole (12):

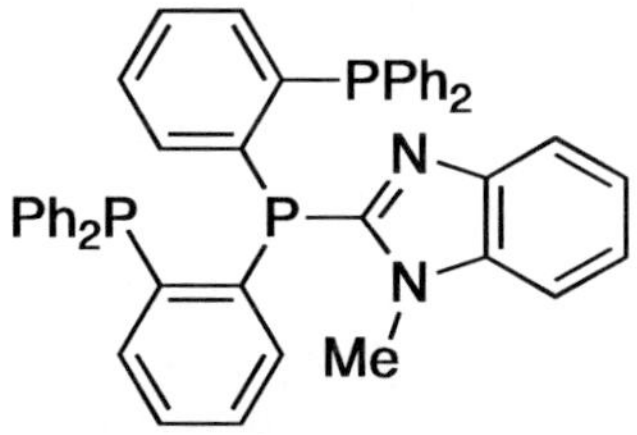

The title compound was obtained as white solid in 62% yield on a 0.5 mmol scale using 1-methylbenzimidazole as the heteroarene. The pure product was obtained by recrystallization from dichloromethane/diethyl ether twice.

Melting point: 163–164 °C (diethyl ether).

^{1}H NMR (500 MHz, $CDCl_3$): δ 7.62 (d, J = 8.1 Hz, 1H), 7.33–7.29 (m, 2H), 7.25–6.98 (m, 29H), 3.46 (s, 3H).

^{31}P NMR (202 MHz, $CDCl_3$): δ −13.0 (d, J = 154 Hz, 2P), −44.6 (t, J = 154 Hz, 1P).

HRMS (APCI +): *m/z* calcd for $C_{44}H_{35}N_2P_3$ [M + H$^+$] 685.2086; found: 685.2090.

2-(Bis(2-(diphenylphosphaneyl)phenyl)phosphaneyl)benzo[*d*]thiazole (13):

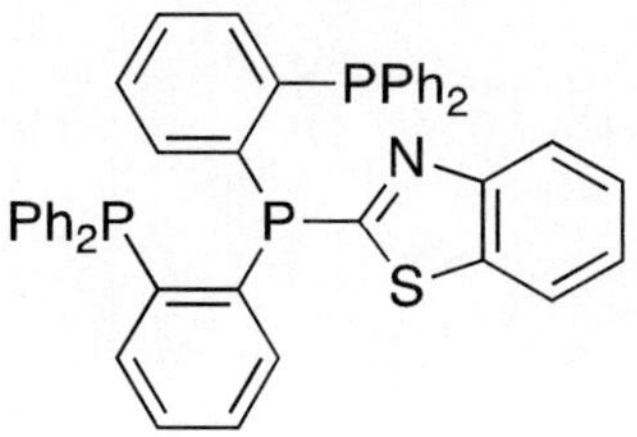

The title compound was obtained as pale yellow solid in 69% yield on a 0.5 mmol scale using benzothiazole as the heteroarene. The pure product was obtained by recrystallization from dichloromethane/methanol twice.

Melting point: 246–248 °C (methanol).

^{1}H NMR (500 MHz, CD_2Cl_2): δ 8.00 (d, J = 8.0 Hz, 1H), 7.82 (d, J = 8.1 Hz, 1H), 7.47–7.02 (m, 30H).

^{31}P NMR (202 MHz, CD_2Cl_2): δ −14.3 (AB_2m, J = 165 Hz, 2P), −20.7 (AB_2m, J = 165 Hz, 1P).

HRMS (APCI +): m/z calcd for $C_{43}H_{32}NP_3S$ [M + H^+] 688.1541; found: 688.1554.

2-Bromo-4,5-dimethylphenyl trifluoromethanesulfonate (15):

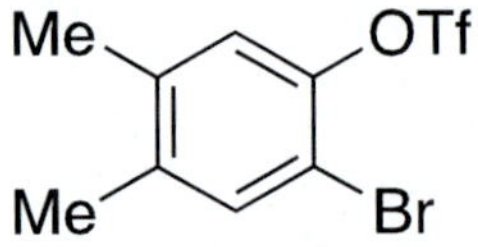

Trifluoromethanesulfonic anhydride (2.7 mL, 17 mmol) was added dropwise at 0 °C to a solution of 2-bromo-4,5-dimethylphenol [13] (3.0 g, 15 mmol) and pyridine (1.8 mL, 23 mmol) in dichloromethane (15 mL). The reaction mixture was gradually warmed to rt and stirred for 12 h. The reaction mixture was quenched by addition of an aqueous solution of HCl (2 M, 5 mL), and the organic layer was washed with water and brine and dried over Na_2SO_4. The solvent was removed under reduced pressure, and the product was used in the next step without further purification.

^{1}H NMR (500 MHz, $CDCl_3$): δ 7.42 (s, 1H), 7.09 (s, 1H), 2.26 (s, 3H), 2.25 (s, 3H).

^{13}C NMR (125 MHz, $CDCl_3$): δ 144.7, 138.8, 138.3, 134.6, 123.4, 118.6 (q, J = 321 Hz), 112.0, 19.6, 19.2.

^{19}F NMR (471 MHz, $CDCl_3$): δ −76.6.

GC MS (EI) m/z (relative intensity): 334 (M^+, 21), 332 (20), 201 (50), 199 (51), 173 (11), 171 (12), 92 (100), 91 (75).

Bis(3,5-dimethylphenyl)phosphine oxide (16):

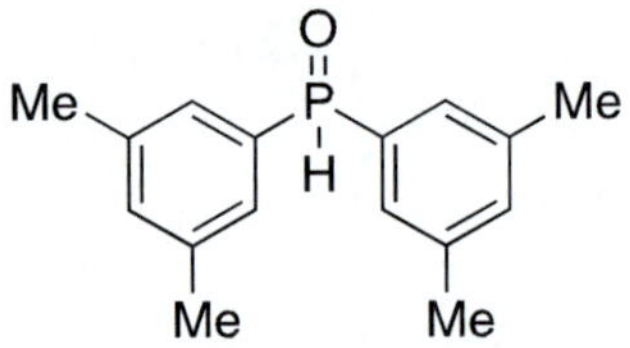

The title compound was prepared according to the literature [14]. The compound data was in good agreement with the literature.

(2-Bromo-4,5-dimethylphenyl)bis(3,5-dimethylphenyl)phosphane (17):

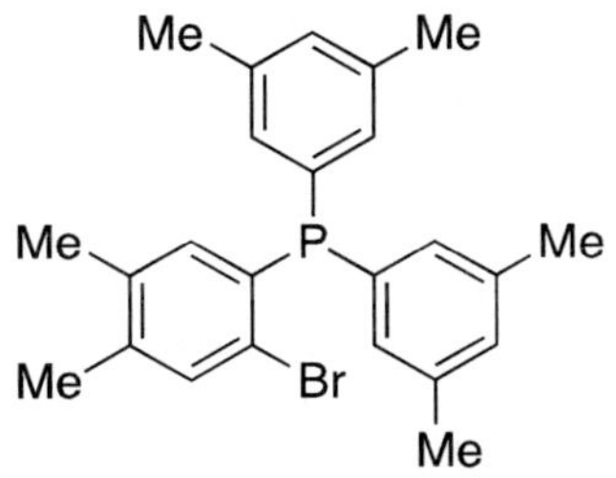

A mixture of 2-bromo-4,5-dimethylphenyl trifluoromethanesulfonate (1.00 g, 3.0 mmol), bis(3,5-dimethylphenyl)phosphine oxide (0.77 g, 3.0 mmol), $Pd_2(dba)_3 \cdot CHCl_3$ (155 mg, 0.15 mmol), 1,3-bis(diphenylphosphino)propane (dppp, 124 mg, 0.30 mmol), and *i*-Pr_2NEt (0.77 mL, 4.5 mmol) in toluene (9.0 mL) was stirred under argon atmosphere at 110 °C for 16 h. The reaction mixture was quenched by addition of an aqueous solution of HCl (1 M, 9 mL), and the aqueous layer was extracted with ethyl acetate (5 mL × 3). The combined organic layers were washed with water and brine and dried over Na_2SO_4. The solvent was removed under reduced pressure and the crude product was purified by silica gel chromatography (hexane: ethyl acetate = 1:1) to afford (2-bromo-4,5-dimethylphenyl)bis(3,5-dimethylphenyl)phosphine oxide containing inseparable impurities (0.75 g in total). This product was used in the next step without further purification.

A mixture of (2-bromo-4,5-dimethylphenyl)bis(3,5-dimethylphenyl)phosphine oxide (0.75 g, 1.7 mmol), trichlorosilane (0.85 mL, 8.5 mmol), and *N*,*N*-dimethylaniline (1.28 mL, 10 mmol) in toluene (11 mL) was stirred under argon atmosphere at 110 °C for 15 h. The reaction mixture was carefully quenched by addition of an aqueous solution of NaOH (25%, 6 mL) at 0 °C and the aqueous layer was extracted with toluene (5 mL × 3). The combined organic layers were washed with an aqueous solution of HCl (1 M), water, and brine, and dried over Na_2SO_4. The solvent was removed under reduced pressure, and the crude product was purified by silica gel chromatography (hexane: ethyl acetate = 50:1) to afford the product as pale yellow solid (0.46 g, 36% over two steps).

Melting point: 120–121 °C (ethyl acetate).

^{1}H NMR (500 MHz, $CDCl_3$): δ 7.37 (d, *J* = 3.7 Hz, 1H), 6.98 (s, 2H), 6.88 (d, *J* = 8.0 Hz, 4H), 6.50 (d, *J* = 2.9 Hz, 1H), 2.27 (s, 12H), 2.23 (s, 3H), 2.06 (s, 3H).

^{31}P NMR (202 MHz, $CDCl_3$): δ −6.1.

HRMS (APCI+): *m/z* calcd for $C_{24}H_{26}BrP$ [$M + H^+$] 427.1011; found: 427.1019.

(((3,5-Dimethylphenyl)phosphanediyl)bis(4,5-dimethyl-2,1-phenylene))bis(bis(3,5-dimethylphenyl)phosphane) (18):

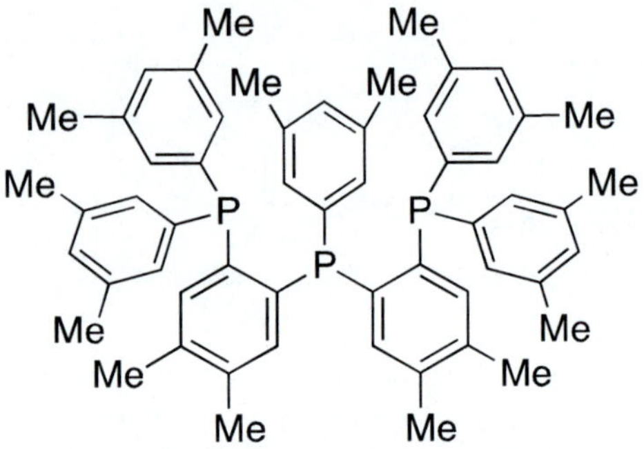

The title compound was obtained as white solid in 48% yield on a 0.25 mmol scale by following the general synthetic procedure for aryl-TPs using (2-Bromo-4,5-dimethylphenyl)bis(3,5-dimethylphenyl)phosphane (**17**) and 1-bromo-3,5-dimethylbenzene as the aryl bromides. The crude reaction mixture was purified by gel permeation chromatography (chloroform).

Melting point: 122–124 °C (chloroform).

^{1}H NMR (500 MHz, $CDCl_3$): δ 6.95–6.93 (m, 2H), 6.83–6.81 (m, 6H), 6.77 (s, 3H), 6.72 (d, $J = 6.6$ Hz, 6H), 6.49–6.47 (m, 2H), 2.14–2.13 (m, 30H), 2.07 (s, 6H), 1.93 (s, 6H).

^{31}P NMR (202 MHz, $CDCl_3$): δ −16.6 (AB_2m, J = 151 Hz, 2P), −20.6 (AB_2m, J = 151 Hz, 1P).

HRMS (APCI +): *m/z* calcd for $C_{56}H_{61}P_3$ [M + H^+] 827.4059; found: 827.4053.

Crystallographic study

The diffraction images for X-ray crystallographic analysis were collected on a Rigaku VariMax Dual equipped with a hybrid photon counting detector using Cu Kα (λ = 1.5418 Å) radiation. A single crystal was coated with mineral oil and mounted on a loop-type mount. The structure was solved by the direct method with SHELXT [15] and refined by the full-matrix least-squares method with SHELXL [16] using Olex2 [17] interface.

Table 3.2Crystal data and structure refinement for benzofuryl-TP (**10**)

Empirical formula	$C_{44}H_{33}OP_3$
Formula weight	670.61
Temperature/K	93
Crystal system	monoclinic

(continued)

(continued)

Empirical formula	$C_{44}H_{33}OP_3$
Space group	$P2_1/n$
a/Å	9.8806(7)
b/Å	21.4720(15)
c/Å	16.9263(11)
$\alpha/°$	90
$\beta/°$	104.235(7)
$\gamma/°$	90
Volume/Å^3	3480.8(4)
Z	4
ρ_{calc} g/cm^3	1.280
μ/mm^{-1}	1.830
F(000)	1400.0
2Θ range for data collection/°	6.78 to 143.606
Index ranges	$-11 \leq h \leq 12, -25 \leq k \leq 23, -20 \leq l \leq 18$
Reflections collected	19,808
Independent reflections	6593 [$R_{int} = 0.0747$, $R_{sigma} = 0.0740$]
Data/restraints/parameters	6593/0/433
Goodness-of-fit on F^2	1.045
Final R indexes [$I \geq 2\sigma(I)$]	$R_1 = 0.0779$, $wR_2 = 0.2120$
Final R indexes [all data]	$R_1 = 0.0924$, $wR_2 = 0.2310$
Largest diff. peak/hole/e Å^{-3}	1.12/−0.54

References

1. (a) Shang R, Ilies L, Nakamura E (2016) J Am Chem Soc 138:10132–10135. (b) Doba T, Ilies L, Sato W, Shang R, Nakamura E (2021) Nat Catal 4:631–638
2. (a) Buscagan TM, Oyala PH, Peters JC (2017) Angew Chem Int Ed 56:6921–6926. (b) Cavaillé A, Joyeux B, Saffon-Merceron N, Nebra N, Fustier-Boutignon M, Mézailles N (2018) Triphos–Fe Chem Commun 54:11953–11956. (c) Schild DJ, Peters JC (2019) ACS Catal 9:4286–4295. (d) Tanabe Y, Nishibayashi Y (2019) Coordination Chem Rev 389:73–93.
3. (a) Chen J-F, Li C (2018) Org Lett 20:6719–6724. (b) Chen J-F, Li C (2020) ACS Catal 10:3881–3889
4. Korstanje TJ, Ivar van der Vlugt J, Elsevier CJ, de Bruin B (2015) Science 350:298–302
5. (a) Adam R, Bheeter CB, Jackstell R, Beller M (2016) ChemCatChem 8:1329–1334. (b) Deng L, Kang B, Englert U, Klankermayer J, Palkovits R (2016) ChemSusChem 9:177–180. (c) Lee HM, Bianchini C, Jia G, Barbaro P (1999) Organometallics 18:1961–1966
6. Liao Q, Saffon N, Mézailles N (2015) ACS Catal 5:6902–6906
7. Zhang J, Duan C, Han C, Yang H, Wei Y, Xu H (2016) Adv Mater 28:5975–5979
8. Keller J, Schlierf C, Nolte C, Mayer P, Straub BF (2006) Synthesis (2):354–365
9. Pirrung MC, Ghorai S (2006) J Am Chem Soc 128:11772–11773

10. Matsumura K, Shimizu H, Saito T, Kumobayashi H (2003) Adv Synth Catal 345:180–184
11. Still WC, Kahn M, Mitra A (1978) J Org Chem 43:2923–2925
12. Pangborn AB, Giardello MA, Grubbs RH, Rosen RK, Timmers FJ (1996) Organometallics 15:1518–1520
13. Cheng X, Peng Y, Wu J, Deng G-J (2016) Org Biomol Chem 14:2819–2823
14. Jin M, Nakamura M (2013) Chem Lett 42:1035–1037
15. Sheldrick GM (2015) Acta Crystallogr Sect A 71:3–8
16. Sheldrick GM (2015) Acta Crystallogr Sect C 71:3–8
17. Dolomanov OV, Bourhis LJ, Gildea RJ, Howard JAK, Puschmann H (2009) J Appl Crystallogr 42:339–341

Chapter 4
Iron-Catalyzed Regioselective Thienyl C–H/C–H Polycondensation

4.1 Introduction

π-Conjugated polymeric compounds have found a wide application in the field of organic electronics such as organic field-effect transistors (OFETs), organic light-emitting diodes (OLEDs), and organic solar cells (OSCs) [1]. Currently, these polymers have been synthesized by a sequence of preinstallation of C–M (M: metal) bond and C–X (X: (pseudo)halide) bond to a hydrocarbon monomer, transition-metal-catalyzed cross-coupling [2], and removal of C–M and C–X bonds by an endcapping procedure [3]. Taking into account the multiple synthetic steps required, transition-metal-catalyzed C–H/C–H polycondensation serves as a straightforward method to synthesize conjugated polymers from simple C–H monomers. However, extremely efficient transition-metal-catalyzed C–H/C–H coupling reaction is essential to achieve polymerization and obtain sufficiently long polymers, which is recognized as a big challenge (Scheme 4.1) [4].

Although direct arylation polymerization (DArP) where the preinstallation of a C–M bond is omitted has been extensively studied since 2010s [5], there are only a limited number of reports on transition-metal-catalyzed C–H/C–H polycondensation. In 2013, Ogino and coworkers reported palladium-catalyzed C–H/C–H polycondensation of simple thiophene monomers to synthesize polythiophenes (Fig. 4.1a) [6]. $Cu(OAc)_2$ and O_2 were used as an oxidant, and CF_3COOH was added to increase the molecular weight of the polymer. NMR analysis of the polymer revealed that there is linkage at the 4-position of thiophene and branching of the polymer chain resulting from low regioselectivity of C–H/C–H coupling (Fig. 4.1b). Later on, it was found that thiophene substrates possessing an ester or an amide group were suitable for polymerization possibly due to the directing effect of a carbonyl group (Scheme 4.2) [7].

In addition to thiophenes, azoles have also been examined as a substrate for transition-metal-catalyzed C–H/C–H polycondensation. In 2014, You and coworkers reported copper-catalyzed C–H/C–H polycondensation of imidazoles [8]. More

T. Doba, *Iron-Catalyzed C–H/C–H Coupling for Synthesis of Functional Small Molecules and Polymers*, Springer Theses,
https://doi.org/10.1007/978-981-99-4121-6_4

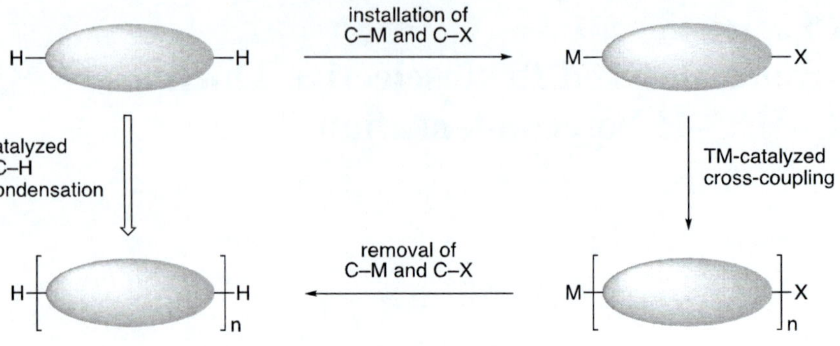

Scheme 4.1 Synthesis of polymer from hydrocarbon monomer

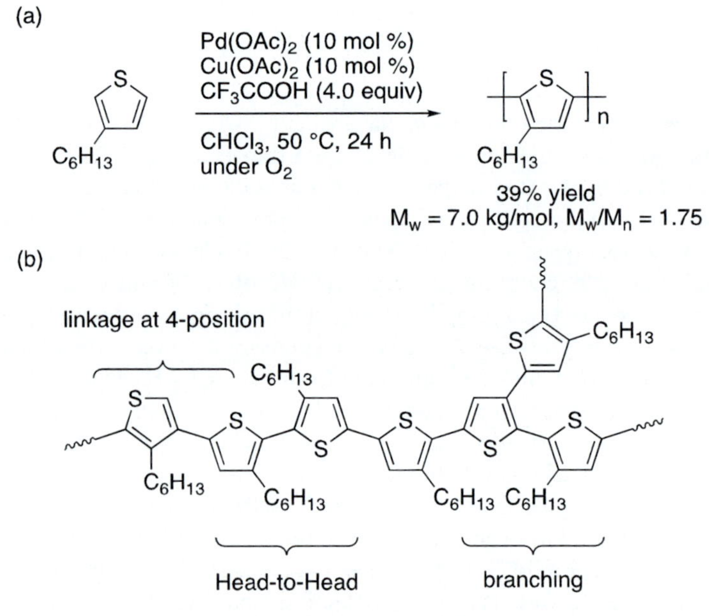

Fig. 4.1 Pioneering work on transition-metal-catalyzed C–H/C–H polycondensation

recently, the same group reported a palladium/copper-cocatalyzed system to achieve C–H/C–H polycondensation of thiazoles (Scheme 4.3) [9].

As explained above, transition-metal-catalyzed C–H/C–H polycondensation has emerged as one of the most straightforward methods to synthesize conjugated polymers. However, this methodology is still underdeveloped and the scope is limited to heteroarenes with special substitution patterns. Also, in all cases, the use of monomers containing electron-rich motifs are avoided to prevent direct oxidation of polymer by external oxidant that is added to turn over the catalyst. In this chapter, I report iron-catalyzed regioselective thienyl C–H/C–H polycondensation that operates under

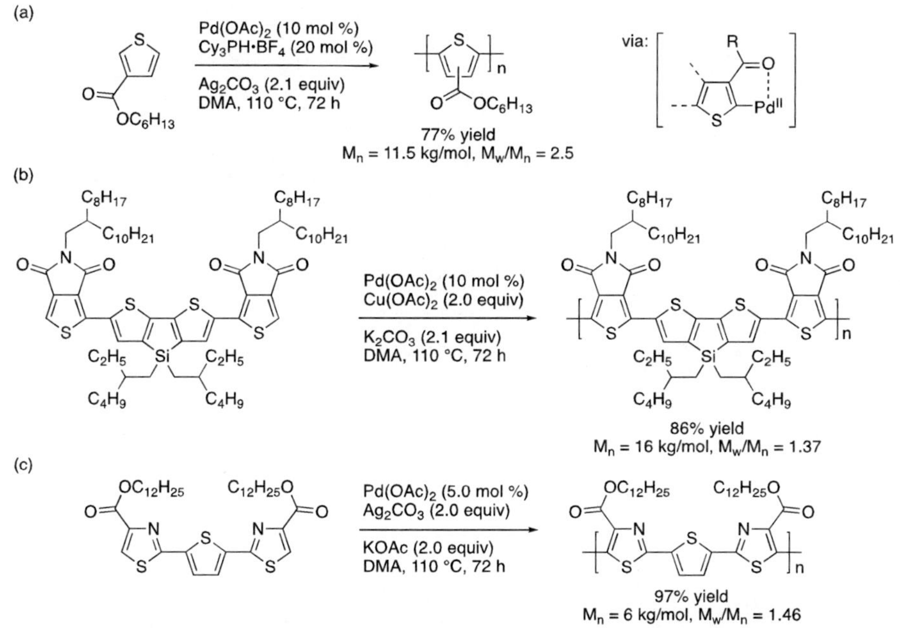

Scheme 4.2 Palladium-catalyzed C–H/C–H polycondensation of carbonyl-containing thiophene monomers

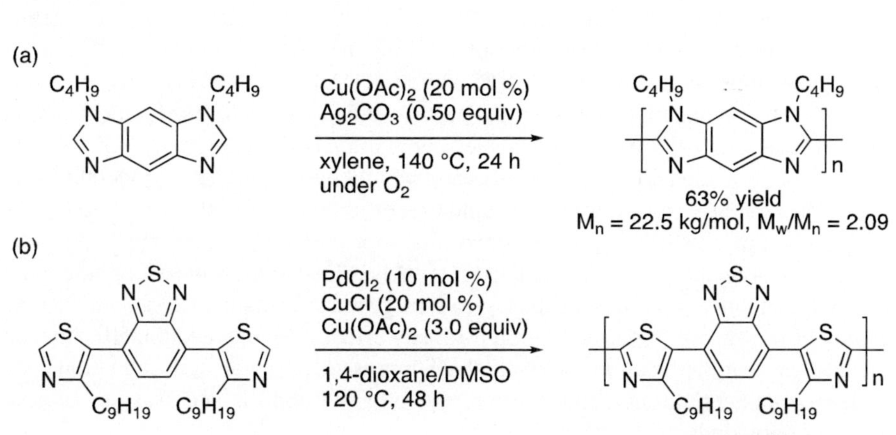

Scheme 4.3 Transition-metal-catalyzed C–H/C–H polycondensation of azole monomers

mildly oxidative conditions and enables polymerization of a wide range of thiophene monomers containing electron-rich and highly conjugated π-motifs. This reaction was realized by improving the efficiency of iron-catalyzed regioselective thienyl C–H/C–H homocoupling by suppression of catalyst deactivation using heteroaryl-TP

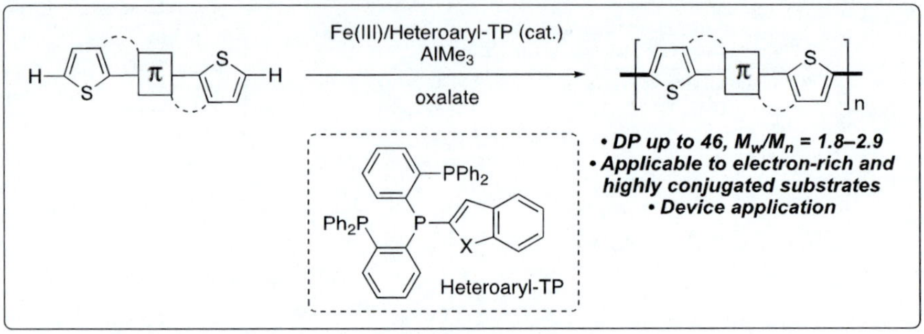

Scheme 4.4 Iron-catalyzed regioselective thienyl C–H/C–H polycondensation

as a ligand (as described in the next section). This method gives direct access to thiophene polymers which are an important class of compounds for optoelectronic device applications (Scheme 4.4).

4.2 Initial Trial

As an initial trial, polycondensation of a dithienylcarbazole monomer (**1**) to poly[(2,7-carbazole)-*alt*-bithiophene] (**2**), which has been reported as a *p*-type semiconductor in OFET [10], was attempted using the reaction conditions for iron-catalyzed regioselective thienyl C–H/C–H homocoupling described in the previous chapter. After quenching the reaction by the addition of acid, the crude reaction mixture was analyzed by analytical gel permeation chromatography (GPC). The number average molecular weight (M_n) and weight average molecular weight (M_w) were determined using polystyrene standards of known molecular weights. Disappointingly, only a short oligomer containing approximately 7 monomer units was obtained. Polydispersity index (PDI, M_w/M_n) was close to the value of 2, indicating that polycondensation proceeds through a step-growth mechanism. Notably, the efficiency of step-growth polymerization is mostly affected by the reaction efficiency at the late stage of the reaction, meaning that suppression of catalyst deactivation even under low concentration of the reactive thienyl C–H bond is necessary to achieve efficient polycondensation (Scheme 4.5).

4.3 Investigation of Iron Source

Assuming that acetylacetonate ($acac^-$) has an inhibition effect by strong binding to iron, the effect of iron source on the efficiency of polymerization was investigated. When $FeCl_3$ was used instead of $Fe(acac)_3$, slight increase of the molecular weight

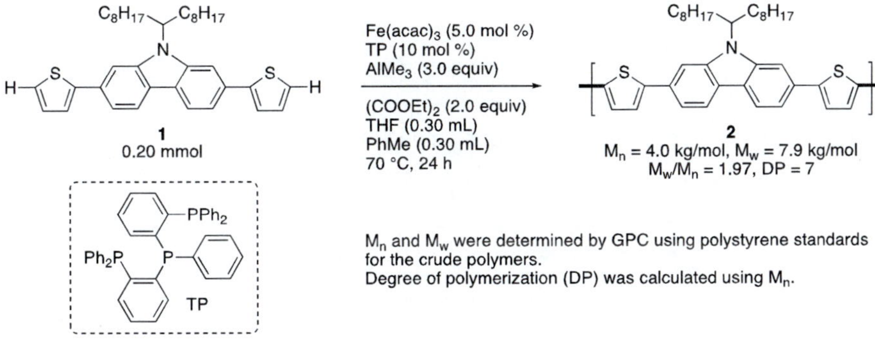

Scheme 4.5 Initial trial of iron-catalyzed regioselective thienyl C–H/C–H polycondensation

was observed (Table 4.1, entry 2). The use of $FeCl_3•6H_2O$, an air stable analog of $FeCl_3$, further improved the reaction efficiency (entry 3). It is noteworthy that 5.0 mol % of $FeCl_3•6H_2O$, releasing 30 mol % of H_2O, had no inhibition effect because of the oxophilicity of Al(III) which prevents O^{2-} binding to Fe(III) catalytic species (entry 2 and 4). Fluorine-containing counteranions were not suitable for the reaction possibly because of the formation of a strong Fe–F bond (entry 5, 6, 8) [11]. Fe(II) salts were less effective than Fe(III) salts, indicating that the reaction proceeds through an Fe(III)/Fe(I) catalytic cycle (entry 7 and 8). Eventually, the DP value of the longest polymer was 9 (entry 3), which prompted me to further improve the efficiency of the iron-catalyzed regioselective thienyl C–H/C–H polycondensation reaction by suppression of catalyst deactivation.

4.4 Determination of Catalyst Deactivation Pathway

To determine the catalyst deactivation pathway at the late stage of the reaction where the concentration of the substrate is significantly low, the outcome of the reaction *without* substrate was considered as an extreme condition. First, $Fe(acac)_3$/TP catalyst, $AlMe_3$, and $(COOEt)_2$ were heated at 70 °C for 3 h without a thiophene substrate and then the substrate was added and heated at 70 °C for 15 h. As expected, the homocoupling product was not obtained at all and the starting material was fully recovered. This clearly demonstrates that the catalyst is totally deactivated in the reaction mixture A (Scheme 4.6).

To understand what happened to the iron catalyst after catalyst deactivation, the tridentate phosphine ligand was recovered from the reaction mixture A by treatment with EDTA. HRMS spectrum of the recovered ligand showed peaks that are heavier than the molecular weight of TP by 14 and 28 (Fig. 4.2). This indicates that the ligand is mono- and di-methylated by the methyl group provided by $AlMe_3$. Although the position of methylation could not be determined by HRMS, it is expected that methylation took place at the *ortho* position of a phenyl group of TP according

Table 4.1 Effect of iron source

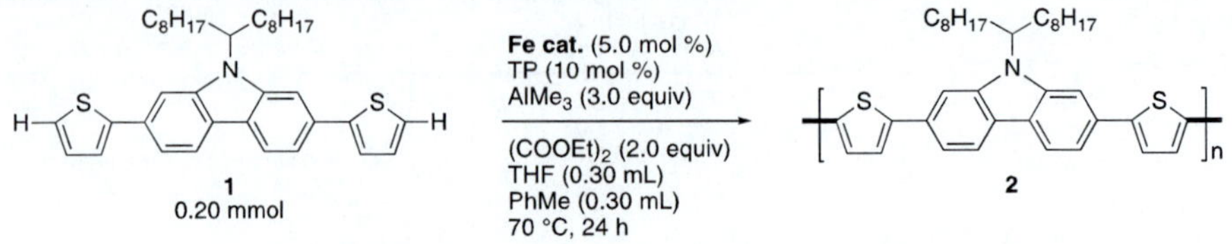

entry	Fe cat.	M_n (kg/mol)[a]	M_w (kg/mol)[a]	M_w/M_n[a]	DP[b]
1	$Fe(acac)_3$	4.0	7.9	1.97	7
2	$FeCl_3$	4.5	9.3	2.06	8
3	$FeCl_3 \cdot 6H_2O$	5.1	10.8	2.12	9
4[c]	$FeCl_3$ + H_2O (30 mol %)	4.6	9.3	2.05	8
5	$FeF_3 \cdot 3H_2O$	2.8	5.0	1.77	5
6	$Fe(OTf)_3$	1.4	1.5	1.10	2
7	$FeCl_2 \cdot 4H_2O$	3.1	5.9	1.88	5
8	$Fe(BF_4)_2 \cdot 6H_2O$	1.8	2.4	1.30	3

[a] M_n and M_w were determined by GPC using polystyrene standards for the crude polymers.
[b] Degree of polymerization was calculated using M_n.
[c] H_2O (30 mol %) was premixed with $AlMe_3$ (3.0 equiv).

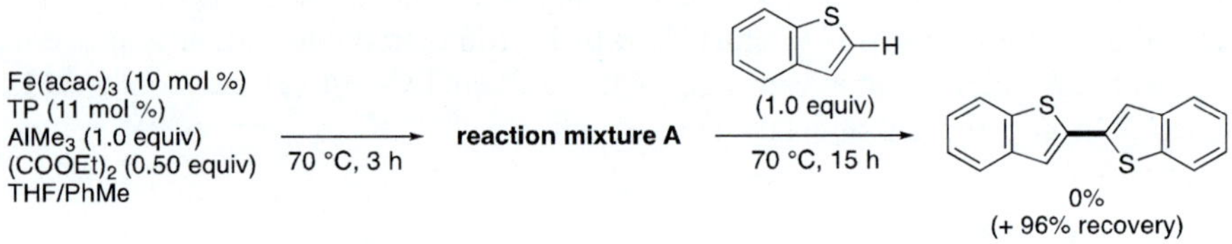

Scheme 4.6 Catalyst deactivation under no substrate

to the precedent of *ortho*-metalation of 1,2-bis(diphenylphosphino)ethane (dppe) under almost identical conditions [12]. First, TP ligand undergoes directed *ortho*-C–H activation intramolecularly through a s-bond metathesis mechanism to form four-membered cyclometalated species **II**. After reductive elimination of the metallacycle with a methyl group, the ligand is methylated (**III**). Because of the compactness of the iron catalytic center, the sterics of the methyl group at the *ortho* position kill the catalytic activity. Therefore, suppression of *ortho*-metalation of the tridentate phosphine ligand would be the key to suppress the catalyst deactivation and achieve efficient polycondensation reaction (Fig. 4.3).

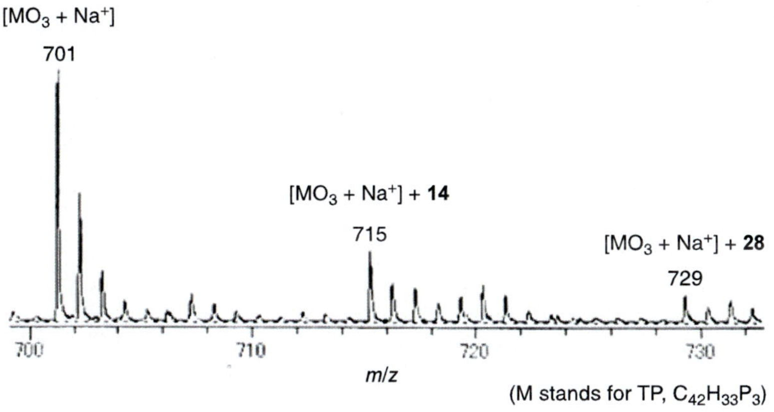

Fig. 4.2 HRMS spectrum of the recovered ligand

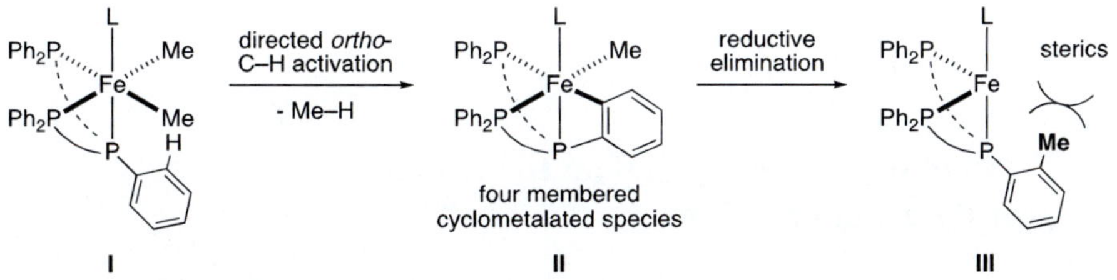

Fig. 4.3 Methylation of TP

4.5 Ligand Design

To suppress *ortho*-metalation of the tridentate phosphine ligand, I designed a new ligand, heteroaryl-TP (Fig. 4.4). The ligand design is based on the assumption that a cyclometalated species of 4-membered ring and 5-membered ring is more strained than that of 4-membered ring and 6-membered ring because of the larger deviation from the ideal bond angles, making it more difficult to form a cyclometalated species and helping to suppress the catalyst deactivation. In other words, as described in the previous chapter, the distance from the central phosphorus atom to the closest hydrogen atom of the central heteroaryl group is over 3.00 Å, which is longer than in the case of Me_2N-TP (2.80 Å), making the directed *ortho*-C–H activation of the ligand more difficult. In addition to this merit, smaller steric hindrance of a heteroaryl group compared to that of a phenyl group may have an effect of facilitating the iron catalysis.

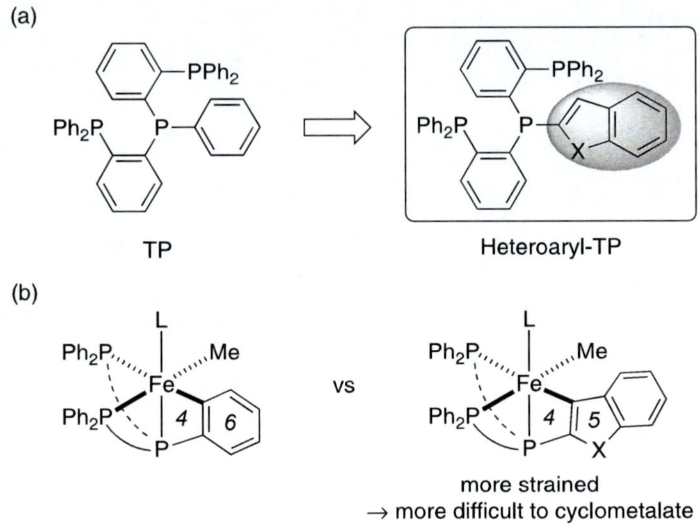

Fig. 4.4 Design of heteroaryl-TP

4.6 Effect of Heteroaryl-TP on the Efficiency of Polycondensation

Heteroaryl-TPs were synthesized using the method described in the previous chapter and used for the polycondensation of dithienylcarbazole monomer (**1**). First, the effect of electronic properties of aryl-TPs was investigated (Table 4.2, entry 1–5). Judging from the molecular weights (M_n and M_w) and the degree of polymerization (DP), installation of an electron-rich aryl group on the central phosphine atom significantly decreased the polymer length. Next, the effect of heteroaryl-TPs on the efficiency of polycondensation was examined. The poor performance of indolyl-TP can be ascribed to the electron-rich indole group and steric hindrance of the *N*-methyl substituent. Benzofuryl- and benzothienyl-TP produced polymer **2** with DP of 23 and 21 with PDI of around 2, which were isolated in 86% and 88% yield, respectively. Notably, branched polymer was not generated as judged by the lack of a conspicuous singlet peak that would appear in the aromatic region of the ^{1}H NMR spectrum in case of the formation of a branched product (Fig. 4.5).

Table 4.2 Effect of ligand on the efficiency of polycondensation

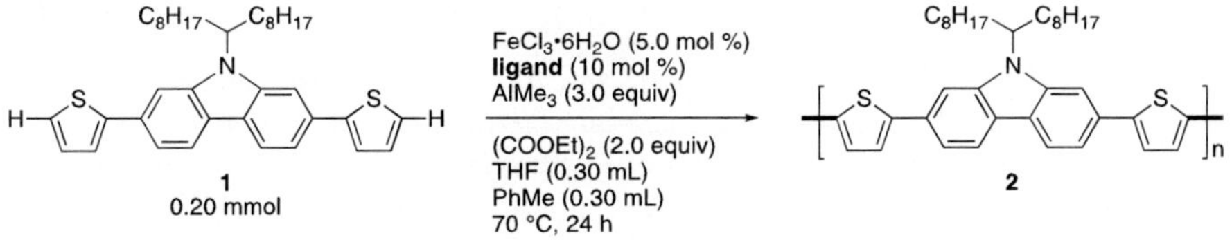

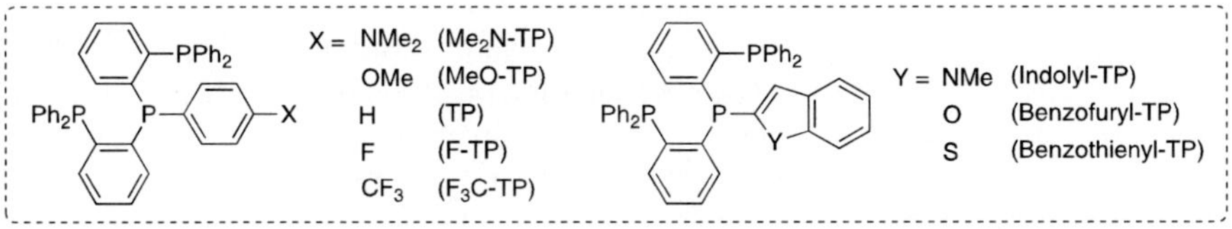

entry	ligand	M_n (kg/mol)[a]	M_w (kg/mol)[a]	M_w/M_n[a]	DP[b]	yield (%)[c]
1	Me_2N-TP	1.6	2.0	1.23	3	-
2	MeO-TP	3.0	5.9	1.99	5	-
3	TP	5.1	10.8	2.12	9	-
4	F-TP	5.0	10.4	2.09	9	-
5	F_3C-TP	6.4	13.2	2.07	11	-
6	Indolyl-TP	1.7	2.2	1.30	3	-
7	Benzofuryl-TP	13.3	22.6	1.70	23	86
8	Benzothienyl-TP	11.9	21.0	1.76	21	88

[a] M_n and M_w were determined by GPC using polystyrene standards for the crude polymers.
[b] Degree of polymerization was calculated using M_n.
[c] Isolated by precipitation. Yields were not determined for short oligomers.

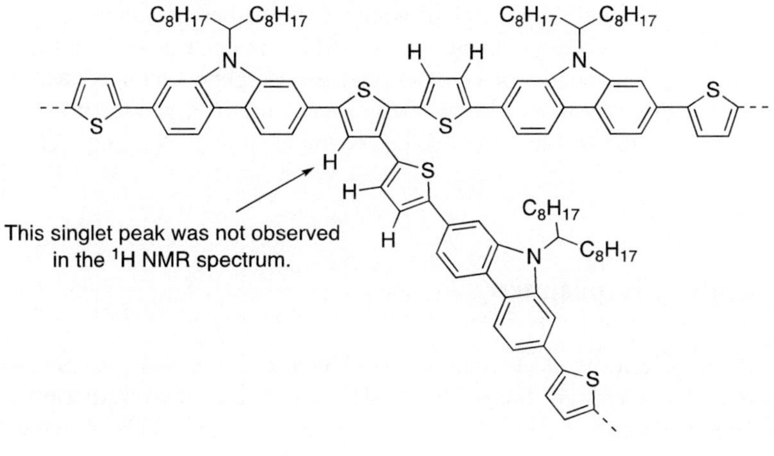

Fig. 4.5 Absence of polymer branching

4.7 Substrate Scope

Using benzofuryl-TP as the optimal ligand, the scope of iron-catalyzed regioselective thienyl C–H/C–H polycondensation was investigated. Table 4.3 shows the examples of straightforward conversion of monomers to the corresponding polymers inaccessible either by the conventional oxidative aromatic C–H/C–H coupling or by palladium-catalyzed C–H/C–H coupling. Homopolymers and copolymers were obtained with DP up to 46 with a unimodal distribution and PDI values of around 2 as confirmed by GPC analyses. The reaction took place exclusively at the C–H bond next to the sulfur atom of the thienyl group, and no branching of the polymers were observed as judged by NMR spectra. This iron-catalyzed method afforded poly(9,9-dioctylfluorene-*alt*-bithiophene) (**3**, abbreviated as F8T2) in 91% yield with a large molecular weight (M_n = 20.3 kg/mol, DP = 37) reaching the length of a commercially available F8T2 (e.g., M_n = 16 kg/mol from Ossila). F8T2 is widely used for organic light-emitting diodes, photovoltaics, and field-effect transistors [13]. Installation of alkyl substituents on monomers for F8T2 synthesis helped to increase M_n to 33 kg/mol and DP to 42–46 (**4**, **5**) by increasing solubility. This method was applicable not only to the synthesis of donor polymers but also to the synthesis of polymers containing electron acceptor moieties such as ester or amide groups (**6**, **7**). The excellent regioselectivity of iron-catalyzed thienyl C–H/C–H coupling allowed us to synthesize a polythiophene (**8**) that contains an average of 69 thiophene units from a 3,3”-dihexyl-2,2’:5’,2”-terthiophene monomer bearing hexyl side chains at predetermined positions [14]. The reaction of a fused thiophene monomer, 4,8-dialkoxybenzo[1,2-*b*:4,5-*b*′]dithiophene, afforded the corresponding polymer with DP of 22 and PDI of 2.16 (**9**). The end-group signals were observed in the ^{1}H NMR spectra of **8** and **9,** and the DP values were determined as 21 and 28, respectively. These values are in good agreement with the DP values determined by GPC analysis using polystyrene standards of known molecular weights. Copolycondensation of two and three different monomers was also successful, in which the initial reactant ratios were reflected to the ratios of the monomer units in the polymer products (**10**, **11**).

4.8 Control Experiments

To prove the superiority of the iron-catalyzed regioselective thienyl C–H/C–H polycondensation to other transition-metal-catalyzed methods, control experiments were conducted. As described above, the iron-catalyzed method produced polymer **2** with DP of 23. However, when palladium-catalyzed methods using either $Cu(OAc)_2$ or Ag_2CO_3 as an oxidant reported by Chen and coworkers[7b,c] were applied to monomer **1**, only short oligomers with DP of 4 or 7 were obtained, respectively. These results clearly show that the iron-catalyzed regioselective thienyl C–H/C–H polycondensation gives straightforward access to polymers inaccessible by previously reported palladium-catalyzed C–H/C–H coupling (Table 4.4).

Table 4.3 Substrate scope

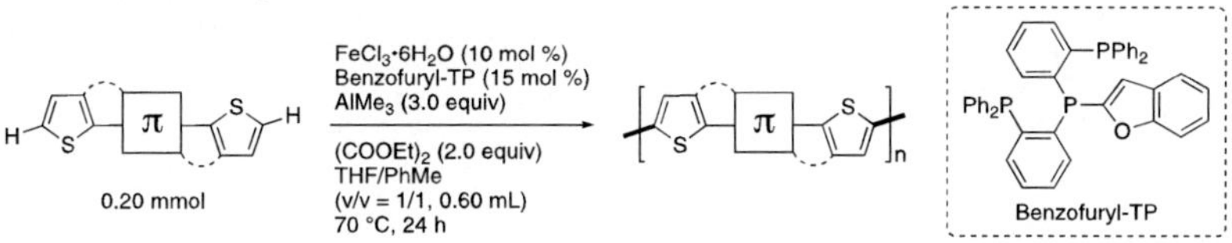

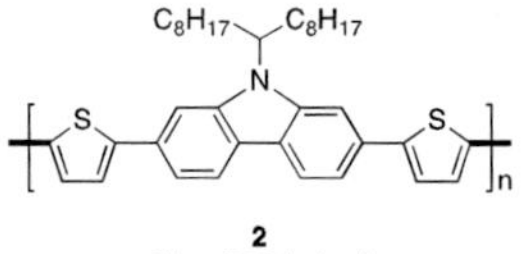

2
M_n = 13.8 kg/mol
M_w = 24.2 kg/mol
M_w/M_n = 1.75
DP = 24
86% yield[a]

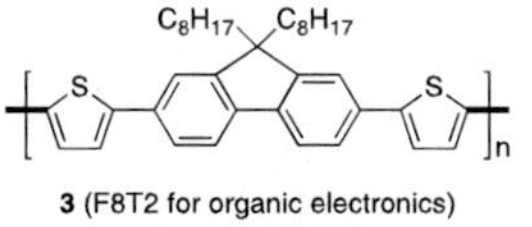

3 (F8T2 for organic electronics)
M_n = 20.3 kg/mol
M_w = 52.9 kg/mol
M_w/M_n = 2.61
DP = 37
91% yield[b,c]

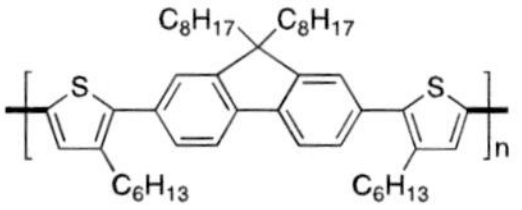

4
M_n = 32.9 kg/mol
M_w = 89.4 kg/mol
M_w/M_n = 2.72
DP = 46
68% yield

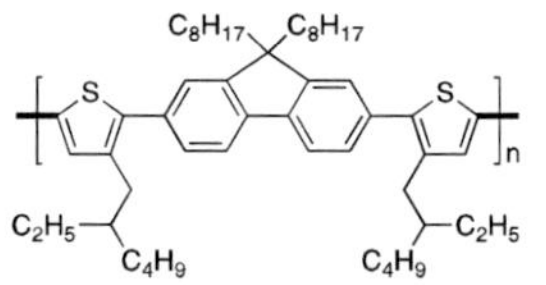

5
M_n = 32.6 kg/mol
M_w = 69.9 kg/mol
M_w/M_n = 2.15
DP = 42
60% yield

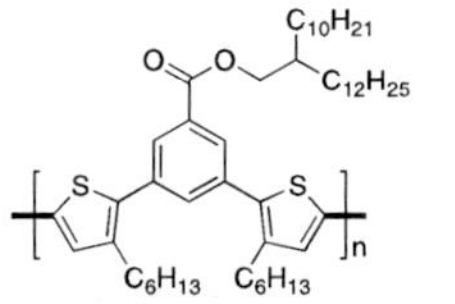

6
M_n = 14.9 kg/mol
M_w = 42.9 kg/mol
M_w/M_n = 2.87
DP = 19
85% yield

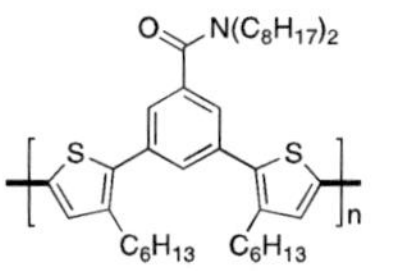

7
M_n = 12.2 kg/mol
M_w = 30.2 kg/mol
M_w/M_n = 2.49
DP = 18
91% yield[c,d]

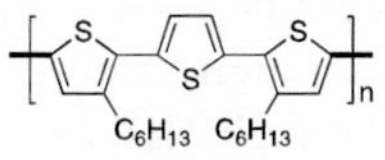

8
M_n = 9.6 kg/mol
M_w = 19.5 kg/mol
M_w/M_n = 2.02
DP = 23
86% yield[e]

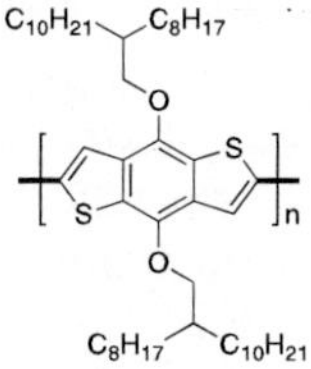

9
M_n = 17.2 kg/mol
M_w = 37.0 kg/mol
M_w/M_n = 2.16
DP = 22
88% yield

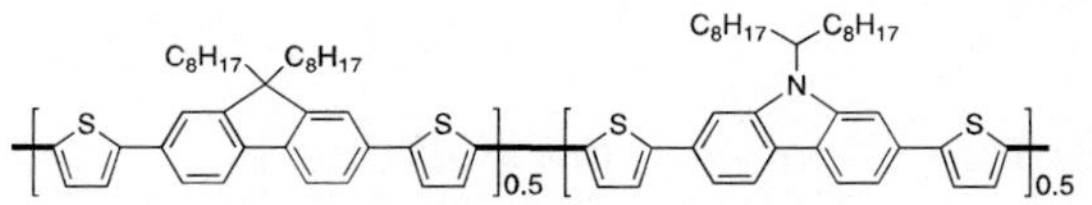

M_n = 19.7 kg/mol, M_w = 38.0 kg/mol
M_w/M_n = 1.93, DP = 35, 98% yield

10

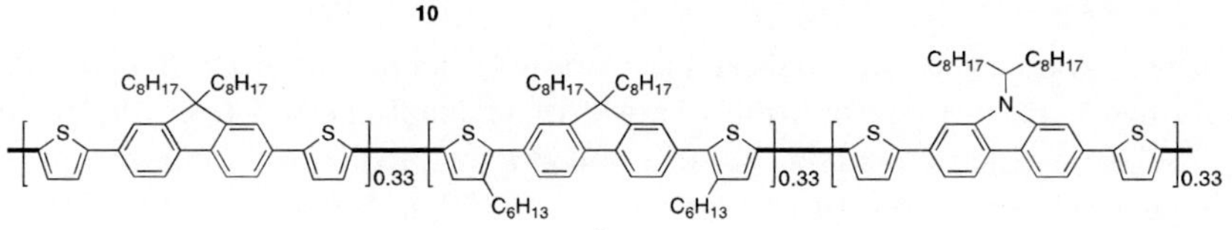

11
M_n = 23.1 kg/mol, M_w = 56.3 kg/mol, M_w/M_n = 2.44, DP = 38, 56% yield

All yields are isolated yields. M_n and M_w were determined by GPC using polystyrene standards for the polymers isolated either by precipitation or by GPC. Degree of polymerization was calculated using M_n. [a] With $FeCl_3·6H_2O$ (5.0 mol %) and benzofuryl-TP (10 mol %). [b] With THF (0.15 mL). [c] For 48 h. [d] With benzofuryl-TP (20 mol %). [e] At 60 °C.

Table 4.4 Control experiments with palladium-catalyzed C–H/C–H coupling

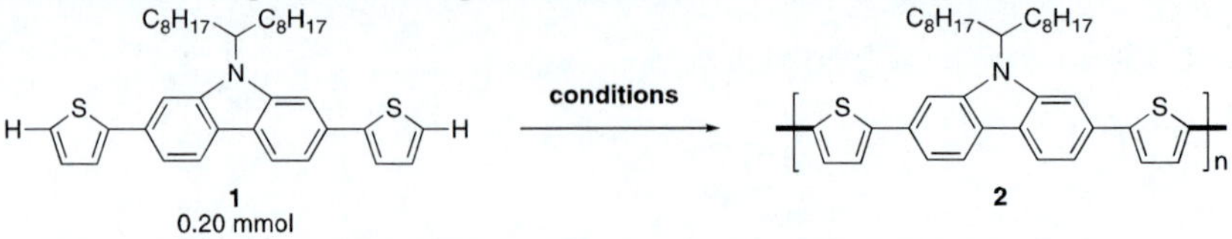

entry	conditions	M_n (kg/mol)[a]	M_w (kg/mol)[a]	M_w/M_n[a]	DP[b]
1	$FeCl_3 \cdot 6H_2O$ (10 mol %), Benzofuryl-TP (15 mol %), $AlMe_3$ (3.0 equiv), $(COOEt)_2$ (2.0 equiv), THF (0.30 mL), PhMe (0.30 mL), 70 °C, 24 h	13.3	22.6	1.70	23
2	$Pd(OAc)_2$ (10 mol %), K_2CO_3 (2.1 equiv), $Cu(OAc)_2$ (2.0 equiv), DMA (2.5 mL), 110 °C, 72 h.	2.0	3.3	1.64	4
3	$Pd(OAc)_2$ (5.0 mol %), KOAc (2.0 equiv), Ag_2CO_3 (2.0 equiv), DMA (2.5 mL), 110 °C, 72 h.	3.7	9.2	2.53	7

[a] M_n and M_w were determined by GPC using polystyrene standards for the crude polymers.
[b] Degree of polymerization was calculated using M_n.

4.9 Mechanism of Polycondensation

To gain insight into the mechanism of iron-catalyzed regioselective thienyl C–H/C–H polycondensation, reaction progress was monitored by sampling the reaction mixture after specific reaction times. Figure 4.6 shows the plots of M_n and M_w/M_n (PDI) as a function of monomer conversion. Late-stage increase of M_n and convergence of PDI to 2 indicated that the reaction proceeds through a step-growth mechanism [15]. This can be ascribed to the weak interaction of the Fe(III) catalyst with the π-surface of the polymer chain (Fig. 4.7a). The polycondensation mechanism of the iron-catalyzed method is in contrast with a chain-growth mechanism where, for example, Pd(0) sticks to the polymer chain through strong interaction with the π-surface and activates the terminus intramolecularly (Fig. 4.7b) [16].

4.10 Removal of Residual Catalyst from Polymer

It has been known that a residual catalyst in the polymer product may affect the device performance [17]. Therefore, the removal of residual catalyst from the polymer synthesized by the iron-catalyzed method was attempted considering the optoelectronic device applications of polymers. Using F8T2 as a model polymer for investigation, it was determined that a thiol-functionalized silica scavenger (purchased from Fuji Silysia Chemical Ltd.) effectively removes both residual iron and phosphorus from the polymer down to 26 ppm and 175 ppm, respectively. The amount of elements were determined by inductively coupled plasma (ICP) analysis. For

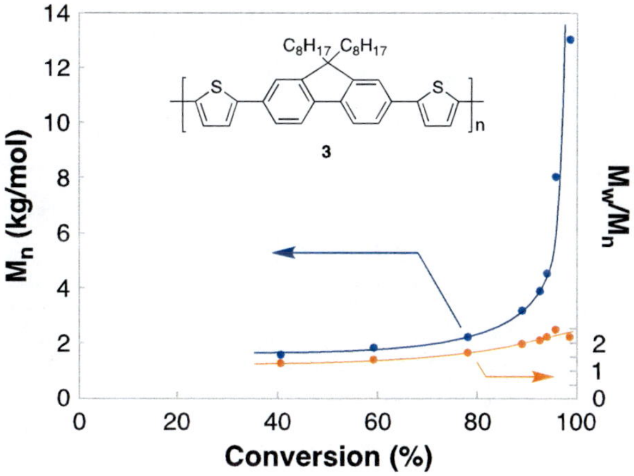

Fig. 4.6 M_n and M_w/M_n of F8T2 as a function of monomer conversion

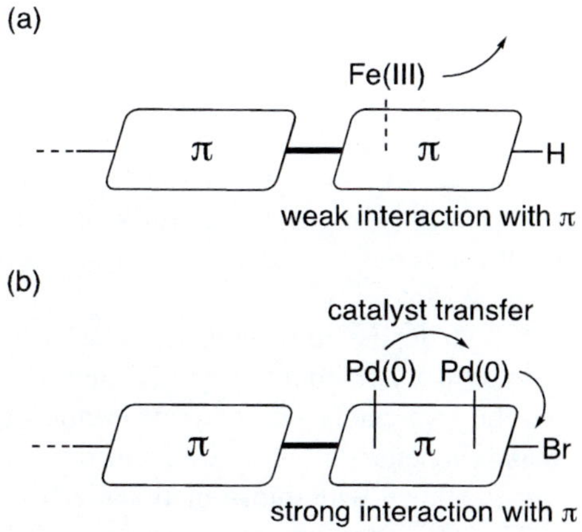

Fig. 4.7 Step-growth polymerization and chain-growth polymerization

comparison, the palladium content of an F8T2 polymer purchased from a commercial supplier was determined to be over 200 ppm, showing the advantage of using iron-catalyzed method for the synthesis of polymer containing less catalyst residues (Fig. 4.8).

4.11 Application to Perovskite Solar Cell

Finally, application of the polymer to optoelectronic devices was attempted. As shown in Fig. 4.9, I designed a new donor polymer (named as Mes-TADHT) that has an enhanced quinoidal conjugation by insertion of P3HT to PTAA, assuming the formation of a stable radical cation in the working device. Retrosynthetic analysis showed that iron-catalyzed C–H/C–H polycondensation of a C–H monomer is superior to transition-metal-catalyzed cross-coupling of dihalide and diorganometallic

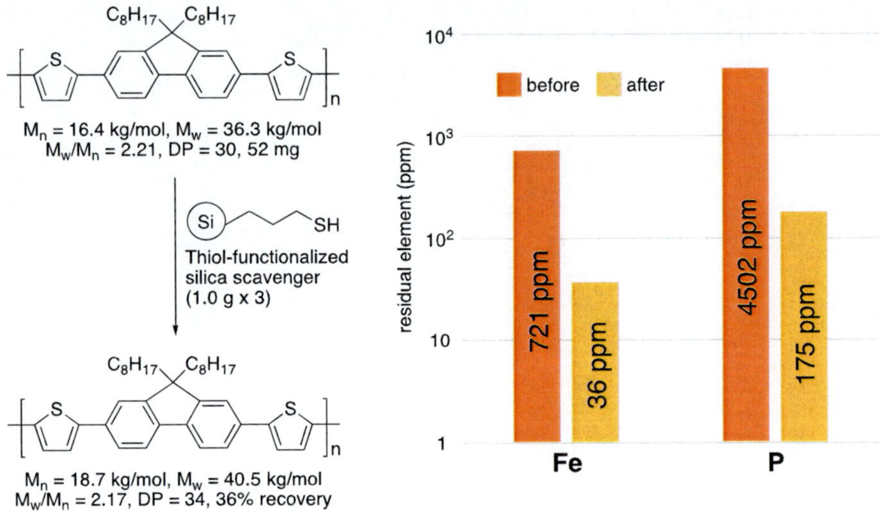

Fig. 4.8 Removal of residual catalyst from F8T2

monomers (Fig. 4.9a, b) not only because it does not require a tedious endcapping procedure but also solves the problems of homocoupling defects caused by transition-metal-catalyzed cross-coupling and strict control of stoichiometry between monomer **A** and **B** [5c].

Based on the retrosynthetic analysis, the C–H monomer was synthesized by Suzuki–Miyaura coupling in one step from a commercially available aryl bromide and the iron-catalyzed polycondensation was conducted using benzofuryl-TP as a ligand (Scheme 4.7). The residual iron content was reduced to less than 600 ppm by post-treatment with thiol-functionalized silica scavenger.

The ionization potential was determined to be 5.4 eV, which perfectly matches with the energy level toward $MAPbI_3$. Therefore, the application of Mes-TADHT as a hole-transporting material (HTM) for a normal structure $MAPbI_3$ perovskite solar cell (PVSC) was attempted in collaboration with Prof. Yutaka Matsuo. All the devices were fabricated by Dr. Hao-sheng Lin in Matsuo Lab. The Mes-TADHT devices doped with 20 wt% of LiTFSI outperformed the PTAA device in terms of open circuit voltage (*V*oc), short circuit current (*J*sc), and fill factor (*FF*). The best Mes-TADHT device performed with a PCE of 21.3%, a V_{OC} of 1.15 V, a J_{SC} of 23.8 mA cm^{-2}, and a *FF* of 0.79. The enhancement of the PCE and V_{OC} value is ascribed to the perfect match of HOMO energy level of the hole-transporting layer with $MAPbI_3$. The stability test of unencapsulated devices using Mes-TADHT and PTAA carried out under ambient atmosphere and humidity showed that the Mes-TADHT device is stable for 1000 h without decay in efficiency, while PTAA device deteriorated significantly. The enhanced stability of the Mes-TADHT device is ascribed to the enhanced hydrophobicity by the hexyl substituents (Fig. 4.10).

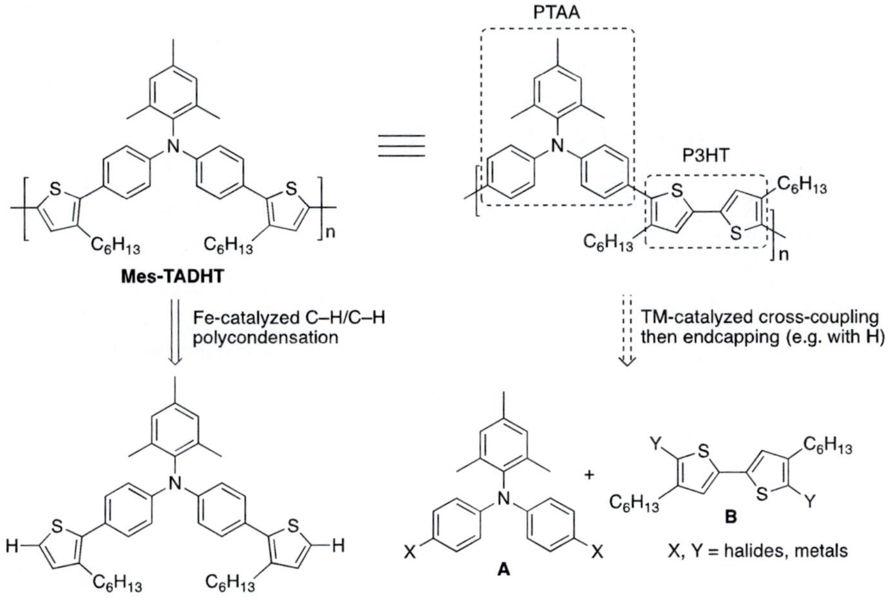

Fig. 4.9 Material design and retrosynthetic analysis

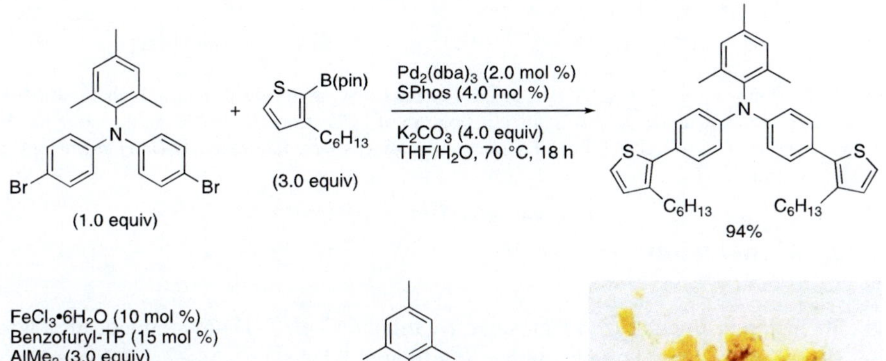

Scheme 4.7 Synthesis of Mes-TADHT

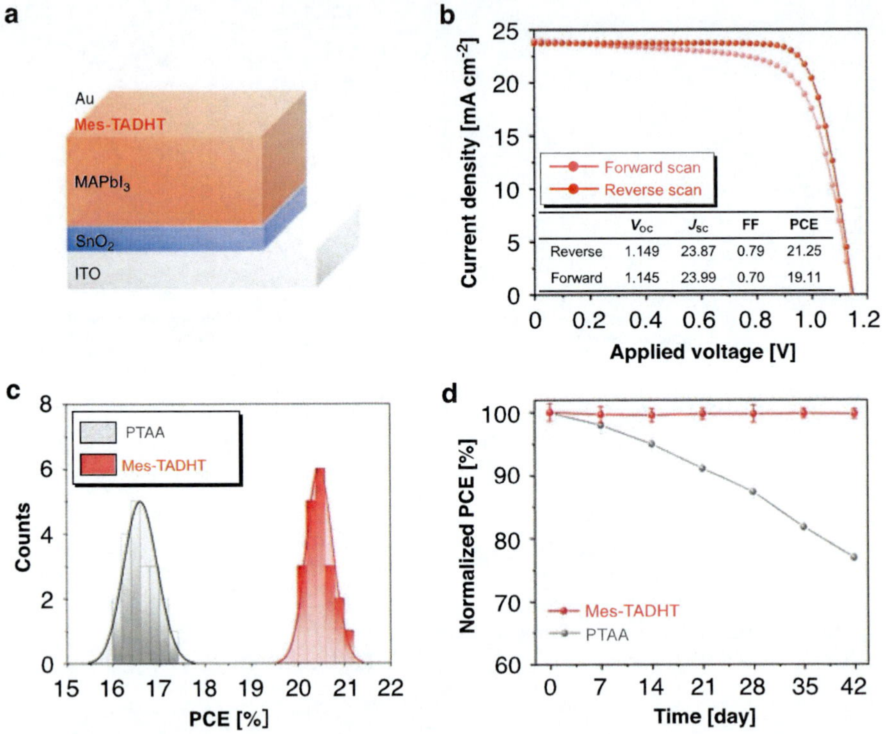

Fig. 4.10 Application of Mes-TADHT as a HTM for PVSC. **a** device structure. **b** champion J–V curves in forward scan and reverse scan. **c** histogram of PCEs from 20 devices in the same batch. **d** stability tests of Mes-TADHT and PTAA devices. Devices were fabricated by Dr. Hao-Sheng Lin in Matsuo Lab

4.12 Conclusion

In conclusion, iron-catalyzed regioselective thienyl C–H/C–H polycondensation was developed using tridentate phosphine as a ligand, $AlMe_3$ as a base, and diketone as a mild oxidant. Intramolecular C–H activation of the ligand was determined as a catalyst deactivation pathway, and the use of heteroaryl-TP was effective to suppress the catalyst deactivation and achieve efficient polycondensation. This reaction took place exclusively at the C–H bond next to the sulfur atom of thienyl group, and no branching of the polymer was observed. Monomers containing various kinds of redox-sensitive π-motifs were well tolerated due to the mildness of the iron catalytic cycle created by the combination of low redox potential of Fe(III)/Fe(I) and mild oxalate oxidant, which enabled the synthesis of a wide range of thiophene materials of interest in materials science. Because of the weak interaction of Fe(III) with the π-surface of the polymer chain, polycondensation proceeded through a step-growth mechanism and the residual catalyst was easily removed by post-treatment with a thiol-functionalized silica scavenger. A hybrid polymer of PTAA and P3HT was synthesized by the iron-catalyzed method and successfully applied as a hole-transporting material for perovskite solar cells in collaboration with Matsuo group. The device showed the highest PCE of 21.3% and a long-term stability over 1000 h, both of which are superior to the state-of-the-art PTAA control device. This work highlights the benefits of iron catalysis for the synthesis of π-conjugated polymeric compounds of importance in energy device applications.

4.13 Experimental

Materials and methods

All air- or moisture-sensitive reactions were performed in a dry reaction vessel under argon atmosphere. Air- or moisture-sensitive liquids and solutions were transferred with syringe or Teflon cannula. The water content of solvents was confirmed to be less than 30 ppm by Karl Fischer titration performed with MKC-210 (Kyoto Electronics Manufacturing Co., Ltd.). Analytical thin-layer chromatography (TLC) was performed with a glass plate coated with 0.25 mm 230–400 mesh silica gel containing a fluorescent indicator. Organic solutions were evacuated with a diaphragm pump through a rotary evaporator. Flash column chromatography was performed as described by Still et al. [18]. Preparative recycling gel permeation chromatography (GPC) was performed with LC-92XX II NEXT instrument (Japan Analytical Industry Co., Ltd.) equipped with JAIGEL-2 h polystyrene columns using chloroform as an eluent at the flow rate of 7.5 mL/min.

Gas chromatography (GC) was performed with GC-2014 instrument (Shimadzu Co.) equipped with an ULBON HR-1 (0.25 mm I.D. × 25 mL, 0.25 mm, Shinwa Chemical Industries, Ltd.) capillary column. Mass spectra (GC–MS) were taken with Parvum 2 instrument (Shimadzu Co.). High-resolution mass spectra (HRMS) were taken with LCMS-IT-TOF (Shimadzu Co.) using reserpine (MW 608.2734) as an internal standard. Analytical GPC was performed with Prominence instrument (Shimadzu Co.) equipped with a KF-805L (Shodex) column at 40 °C using chloroform as an eluent at the flow rate of 1.0 mL/min. Calibration curves were obtained with ReadyCal Kit (Polymer Standards Service, GmbH) standard polystyrenes. Melting points of solid compounds were measured on a Mel-Temp capillary melting-point apparatus and were uncorrected. Nuclear magnetic resonance (NMR) spectra were taken with ECZ-500 (JEOL, Ltd.) at room temperature unless otherwise noted and reported in parts per million (ppm). ^{1}H NMR spectra were internally referenced to tetramethylsilane (0.00 ppm), $CHCl_3$ (7.26 ppm), $CHDCl_2$ (5.32 ppm), C_2HDCl_4 (5.97 ppm), or $(CHD_2)(CD_3)SO$ (2.50 ppm). ^{13}C NMR spectra were internally referenced to tetramethylsilane (0.0 ppm), $CDCl_3$ (77.0 ppm), CD_2Cl_2 (53.8 ppm), $C_2D_2Cl_4$ (73.8 ppm), or $(CD_3)_2SO$ (39.5 ppm). ^{19}F NMR spectra were internally referenced to C_6F_6 (–164.9 ppm). ^{31}P NMR spectra were internally referenced to $(CH_3O)_3PO$ (2.1 ppm). ICP analysis was performed on Shimadzu ICPS-7510 equipment.

Unless otherwise noted, reagents were purchased from Tokyo Chemical Industry Co., Ltd., Sigma-Aldrich Co., LCC, FUJIFILM Wako Pure Chemical Co., and other commercial suppliers and were used as received. Anhydrous tetrahydrofuran and diethyl ether were purchased from KANTO Chemical Co., Inc. and purified prior to use by a solvent purification system (GlassContour) equipped with columns of activated alumina and supported copper catalyst [19]. $Fe(acac)_3$ (99.9% trace metal basis) was purchased from Sigma-Aldrich Co., LCC and used as received. $FeCl_3{\bullet}6H_2O$ (99.0 + %) was purchased from FUJIFILM Wako Pure Chemical Co. and used as received. Diethyl oxalate was purchased from Tokyo Chemical Industry Co., degassed by Freeze–Pump–Thaw cycling for three times, dried with molecular sieves 4A and kept in a storage flask. Concentrated sulfuric acid (for trace analysis), nitric acid 1.42 (for trace analysis), chloroform (primepure), and methanol (primepure) for ICP analysis were purchased from KANTO Chemical Co., Inc. and used as received. Stating materials were synthesized according to the literature [20].

A representative procedure for the investigation of key reaction parameters of polycondensation (Tables 4.1 and 4.2).

In an oven-dried Schlenk tube was added 9-(heptadecan-9-yl)-2,7-di(thiophen-2-yl)-9*H*-carbazole (114 mg, 0.20 mmol) and benzofuryl-TP (13 mg, 0.020 mmol), and the Schlenk tube was evacuated and refilled with argon for three times. Then a THF solution of $FeCl_3{\bullet}6H_2O$ (0.033 mol/L, 0.30 mL, 0.010 mmol) was added, and a toluene solution of $AlMe_3$ (2.0 mol/L, 0.30 mL, 0.60 mmol) was added by rinsing the wall of the Schlenk tube at rt. Finally, diethyl oxalate (54 μ L, 0.40 mmol) was added and the reaction mixture was degassed by Freeze–Pump–Thaw cycling for three times. After stirring with an agitation rate of 500 rpm at 70 °C for 24 h, the reaction mixture was diluted with THF (2 mL) and quenched by addition of a dioxane solution of HCl (4 mol/L, 0.50 mL), and stirred for 1 h. ~10 mg of the resulting dark brown solution was sampled, diluted with 1 mL of chloroform and analyzed by GPC to determine the M_n, M_w, M_w/M_n values. Degree of polymerization (DP) was determined by dividing the M_n value by the molecular weight of the monomer unit. The rest of the crude solution was transferred to a 15 mL centrifuge tube and further diluted with THF up to 5 mL. Water was added up to 10 mL, and the tube was shaken vigorously after which orange polymer precipitated immediately. The mixture was centrifuged (4000 rpm, 10 min), and the solution was discarded. Then THF/water (1:1) was added up to 10 mL and the tube was shaken vigorously and centrifuged (4000 rpm, 10 min), and the solution was discarded. Finally, methanol was added up to 10 mL and the tube was shaken vigorously and the solid was filtered. The orange polymer was dried under vacuum overnight.

Note: The reaction mixture remained homogeneous over the entire course of reaction, and we observed little effects of stirring rate on the outcome of the reaction. Technically, too fast stirring is not recommended because the magnetic stirring bar cannot catch up with the stirring rate of the stirrer causing vibration or jump of the stirring bar when the reaction mixture became viscous as polymer accumulates in solution.

A general procedure for iron-catalyzed thienyl C–H/C–H polycondensation (Table 4.3)

Procedure A: In an oven-dried Schlenk tube was added a monomer (0.20 mmol) and benzofuryl-TP (20 mg, 0.030 mmol), and the Schlenk tube was evacuated and refilled with argon for three times. Then a THF solution of $FeCl_3•6H_2O$ (0.067 mol/L, 0.30 mL, 0.020 mmol) was added, and a toluene solution of $AlMe_3$ (2.0 mol/L, 0.30 mL, 0.60 mmol) was added by rinsing the wall of the Schlenk tube at rt. Finally, diethyl oxalate (54 μ L, 0.40 mmol) was added and the reaction mixture was degassed by Freeze–Pump–Thaw cycling for three times. After stirring with an agitation rate of 500 rpm at 70 °C for 24 h, the reaction mixture was diluted with THF (2 mL), quenched by addition of a dioxane solution of HCl (4 mol/L, 0.50 mL), and stirred for 1 h. The crude solution was transferred to a 15 mL centrifuge tube and further diluted with THF up to 5 mL. Water was added up to 10 mL, and the tube was shaken vigorously after which a solid polymer precipitated immediately. The mixture was centrifuged (4000 rpm, 10 min), and the solution was discarded. Then THF/water (1:1) was added up to 10 mL and the tube was shaken vigorously and centrifuged (4000 rpm, 10 min), and the solution was discarded. Finally, methanol was added up to 10 mL and the tube was shaken vigorously. The solid was filtered and dried under vacuum overnight. A chloroform solution of the product was prepared and analyzed by GPC to determine the M_n, M_w, M_w/M_n values of the isolated polymer. Degree of polymerization (DP) was determined by dividing the M_n value by the molecular weight of the monomer unit.

Procedure B: In an oven-dried Schlenk tube was added a monomer (0.20 mmol) and benzofuryl-TP (20 mg, 0.030 mmol), and the Schlenk tube was evacuated and refilled with argon for three times. Then a THF solution of $FeCl_3•6H_2O$ (0.067 mol/L, 0.30 mL, 0.020 mmol) was added, and a toluene solution of $AlMe_3$ (2.0 mol/L, 0.30 mL, 0.60 mmol) was added by rinsing the wall of the Schlenk tube at rt. Finally, diethyl oxalate (54 μ L, 0.40 mmol) was added and the reaction mixture was degassed by Freeze–Pump–Thaw cycling for three times. After stirring with an agitation rate of 500 rpm at 70 °C for 24 h, the reaction mixture was diluted with THF (2 mL), quenched by addition of a dioxane solution of HCl (4 mol/L, 0.50 mL), and stirred for 1 h. The crude solution was diluted with chloroform (10 mL) and washed with water. When emulsion occurred, the solution was filtered through a membrane filter to give a clear solution. The organic phase was dried over Na_2SO_4 and passed through a pad of Florisil. The solvent was removed under reduced pressure, and the crude product was purified by gel permeation chromatography (chloroform) to afford the product. The product was dried under vacuum overnight. A chloroform solution of the product was prepared and analyzed by GPC to determine the M_n, M_w, M_w/M_n values of the isolated polymer. Degree of polymerization (DP) was determined by dividing the M_n value by the molecular weight of the monomer unit.

Polymer 2:

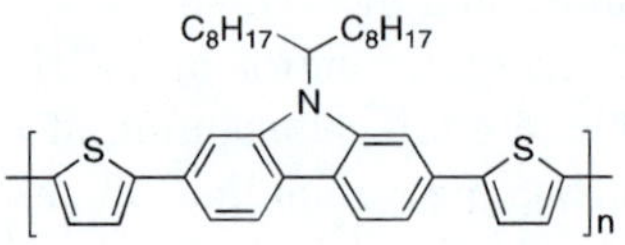

The title compound was obtained as orange solid in 86% yield. The reaction was performed according to Procedure A using benzofuryl-TP (13 mg, 0.020 mmol) and a THF solution of $FeCl_3 \cdot 6H_2O$ (0.033 mol/L, 0.30 mL, 0.010 mmol). This compound is known.[10]

^{1}H NMR (500 MHz, $C_2D_2Cl_4$, 120 °C): δ 8.10 (br d, J = 8.2 Hz, 2H), 7.75 (br s, 2H), 7.55 (br d, J = 8.2 Hz, 2H), 7.38 (br d, J = 3.1 Hz, 2H), 7.30 (br d, J = 3.1 Hz, 2H), 4.75–4.60 (br, 1H), 2.48–2.38 (br, 2H), 2.20–2.05 (br, 2H), 1.38–1.25 (br, 24H), 0.88 (br t, J = 7.2 Hz, 6H).

^{13}C NMR (125 MHz, $C_2D_2Cl_4$, 120 °C): δ 144.8, 136.6, 131.7, 124.9, 124.5, 123.6, 122.8, 120.3, 117.5, 107.4, 56.7, 33.8, 31.5, 29.1, 29.0, 28.8, 26.7, 22.2, 13.6.

M_n = 13.8 kg/mol, M_w = 24.2 kg/mol, M_w/M_n = 1.75, DP = 24.

Anal. Calcd for $(C_{37}H_{45}NS_2)_n$: C, 78.26; H, 7.99; N, 2.47. Found: C, 77.62; H, 7.53; N, 2.33.

Polymer 3:

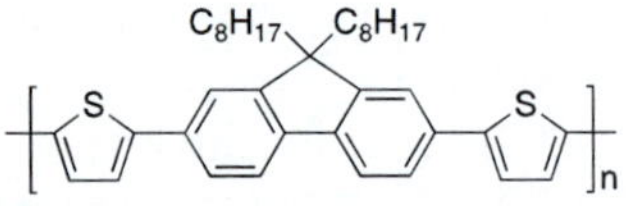

The title compound was obtained as yellow-orange solid in 91% yield. The reaction was performed according to Procedure A using THF (0.15 mL), and the reaction time was 48 h. The compound data was in good agreement with that of commercially available F8T2 (Sigma-Aldrich Co., LCC).

^{1}H NMR (500 MHz, $C_2D_2Cl_4$, 120 °C): δ 7.46 (br d, J = 8.1 Hz, 2H), 7.37–7.35 (br, 4H), 7.10–7.03 (br, 2H), 7.00–6.95 (br, 2H), 1.95–1.75 (br, 4H), 0.99–0.89 (br, 20H), 0.58 (br t, J = 7.2 Hz, 10H).

^{13}C NMR (125 MHz, $C_2D_2Cl_4$, 120 °C): δ 151.9, 144.2, 140.3, 136.5, 132.9, 124.7, 124.5, 123.5, 120.1, 119.9, 55.3, 39.9, 31.5, 29.7, 28.9, 28.8, 23.8, 22.2, 13.6.

M_n = 20.3 kg/mol, M_w = 52.9 kg/mol, M_w/M_n = 2.61, DP = 37.

Polymer 4:

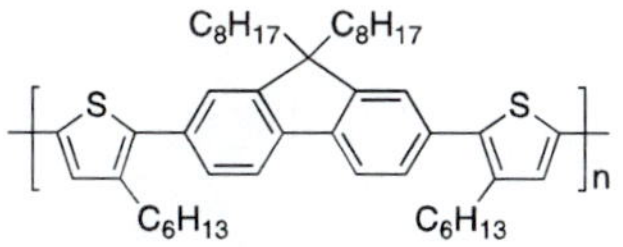

The title compound was obtained as yellow solid in 68% yield. The reaction was performed according to Procedure B.

^{1}H NMR (500 MHz, $C_2D_2Cl_4$, 120 °C): δ 7.65 (br d, J = 8.1 Hz, 2H), 7.38–7.36 (br, 4H), 7.00 (br s, 2H), 2.65–2.58 (br, 4H), 2.05–1.80 (br, 4H), 1.65–1.55 (br, 4H), 1.30–1.00 (br, 32H), 0.82–0.74 (br, 16H).

^{13}C NMR (125 MHz, $C_2D_2Cl_4$, 120 °C): δ 151.3, 139.9, 139.4, 137.5, 135.2, 133.2, 128.0, 126.1, 123.7, 119.5, 55.1, 40.0, 31.5, 31.4, 30.5, 29.8, 29.0 (two signals overlapped), 28.9, 28.8, 23.9, 22.2 (two signals overlapped), 13.6 (two signals overlapped).

M_n = 32.9 kg/mol, M_w = 89.4 kg/mol, M_w/M_n = 2.72, DP = 46.

Polymer 5:

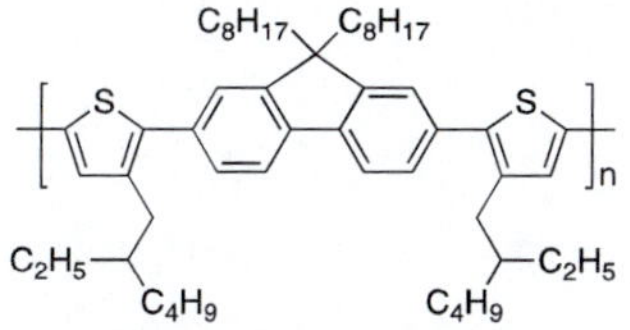

The title compound was obtained as yellow solid in 60% yield. The reaction was performed according to Procedure B.

^{1}H NMR (500 MHz, $C_2D_2Cl_4$, 120 °C): δ 7.64 (br d, J = 8.3 Hz, 2H), 7.37–7.36 (br, 4H), 6.98 (br s, 2H), 2.60–2.55 (br, 4H), 2.05–1.95 (br, 4H), 1.55–1.50 (br, 2H), 1.35–1.05 (br, 36H), 0.79–0.71 (br, 22H).

^{13}C NMR (125 MHz, $C_2D_2Cl_4$, 120 °C): δ 151.3, 140.0, 138.5, 138.1, 135.2, 133.4, 128.3, 126.4 (multiple signals were observed because of slow rotation of bonds), 126.0 (multiple signals were observed because of slow rotation of bonds), 124.0, 119.5, 55.2, 40.6 (multiple signals were observed because of slow rotation of bonds), 40.4 (multiple signals were observed because of slow rotation of bonds), 40.1, 33.3, 32.8, 31.5, 29.9, 29.0, 28.9, 28.7, 26.0, 23.9, 22.7, 22.2, 13.6 (two signals overlapped), 10.5.

M_n = 32.6 kg/mol, M_w = 69.9 kg/mol, M_w/M_n = 2.15, DP = 42.

Polymer 6:

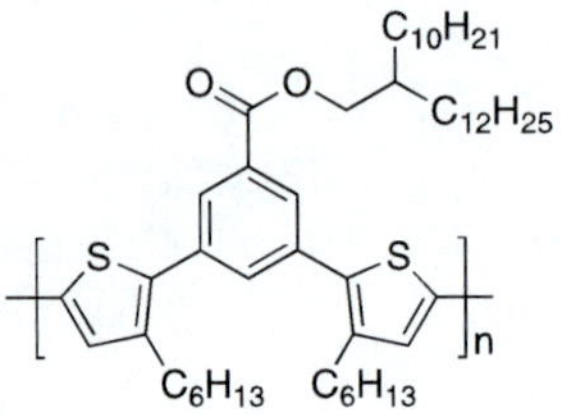

The title compound was obtained as brown solid in 85% yield. The reaction was performed according to Procedure B. Small peaks observed in the ^{1}H NMR spectrum (~5% to the main peaks) were assigned to the monomer unit that underwent transesterification with ethanol of diethyl oxalate.

^{1}H NMR (500 MHz, $C_2D_2Cl_4$, 120 °C): δ 8.02 (br s, 2H), 7.67 (br s, 1H), 7.06 (br s, 2H), 4.25–4.15 (br, 2H), 2.70–2.60 (br, 4H), 1.75–1.70 (br, 1H), 1.70–1.60 (br, 4H), 1.40–1.10 (br, 52H), 0.80–0.75 (br, 12H).

^{13}C NMR (125 MHz, $C_2D_2Cl_4$, 120 °C): δ 166.0, 140.4, 135.7, 135.0, 134.9, 133.3, 131.2, 128.6, 126.2, 68.1, 37.3, 31.8, 31.8, 31.5 (two signals overlapped), 31.3, 30.7, 29.9, 29.6, 29.5, 29.5, 29.5, 29.2, 29.2, 29.1, 28.8, 26.7, 22.6 (two signals overlapped), 22.5, 14.1, 14.0 (two signals overlapped).

$M_n = 14.9$ kg/mol, $M_w = 42.9$ kg/mol, $M_w/M_n = 2.87$, DP = 19.

Polymer 7:

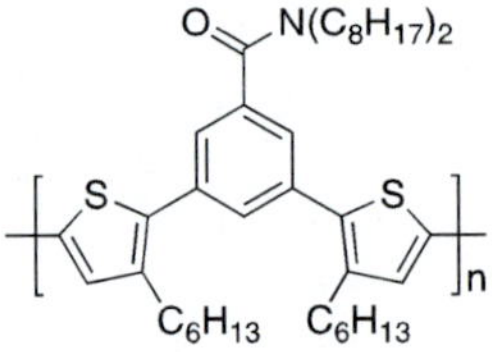

The title compound was obtained as brown solid in 91% yield. The reaction was performed according to Procedure B using benzofuryl-TP (27 mg, 0.040 mmol), and the reaction time was 48 h.

^{1}H NMR (500 MHz, $C_2D_2Cl_4$, 120 °C): δ 7.50 (br s, 1H), 7.31 (br s, 2H), 6.98 (br s, 2H), 3.35–3.25 (br, 4H), 2.65–2.55 (br, 4H), 1.60–1.45 (br, 8H), 1.35–1.10 (br, 32H), 0.80–0.75 (br t, $J = 7.2$ Hz, 12H).

^{13}C NMR (125 MHz, $C_2D_2Cl_4$, 120 °C): δ 170.3, 140.2, 137.9, 135.6, 135.1, 134.7, 129.9, 126.1, 125.8, 123.4, 120.2, 99.3, 49.1, 44.7, 31.7, 31.6, 31.6, 30.8, 29.3, 29.2,

29.1, 29.1, 28.9, 28.7, 27.4, 27.0, 26.5, 22.6, 22.5, 14.1, 14.1. (Multiple signals were observed because of slow rotation of bonds.)

$M_n = 12.2$ kg/mol, $M_w = 30.2$ kg/mol, $M_w/M_n = 2.49$, DP = 18.

Polymer 8:

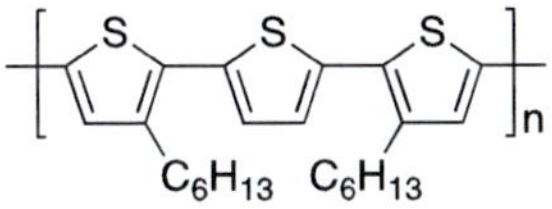

In an oven-dried Schlenk tube was added 3,3''-dihexyl-2,2':5',2''-terthiophene (83 mg, 0.20 mmol) and benzofuryl-TP (20 mg, 0.030 mmol), and the Schlenk tube was evacuated and refilled with argon for three times. Then a THF solution of $FeCl_3$•$6H_2O$ (0.067 mol/L, 0.30 mL, 0.020 mmol) was added, and a toluene solution of $AlMe_3$ (2.0 mol/L, 0.30 mL, 0.60 mmol) was added by rinsing the wall of the Schlenk tube at rt. Finally, diethyl oxalate (54 μ L, 0.40 mmol) was added and the reaction mixture was degassed by Freeze–Pump–Thaw cycling for three times. After stirring with an agitation rate of 500 rpm at 60 °C for 24 h, the reaction mixture was diluted with THF (2 mL), quenched by addition of a dioxane solution of HCl (4 mol/L, 0.50 mL), and stirred for 1 h. The crude solution was diluted with chloroform (10 mL), and the insoluble purple solid was filtered.

The solid was washed with minimal amounts of chloroform, THF, an aqueous solution of HCl (1 M), and acetone to afford the product as purple solid (71 mg, 86%). This compound is known [21]. The product was dissolved in $C_2D_2Cl_4$ at 120 °C, diluted with chloroform, and the soluble portion was analyzed by GPC to determine the M_n, M_w, M_w/M_n values. Small peaks observed in the ^{1}H NMR spectrum were assigned to the thiophene termini of the polymer. By integration of these peaks, DP was calculated to be $4.000/(0.096 + 0.105) + 1 = 21$.

^{1}H NMR (500 MHz, $C_2D_2Cl_4$, 120 °C): δ 7.14 (br s, 2H), 7.06 (br s, 2H), 2.90–2.85 (br, 4H), 1.80–1.75 (br, 4H), 1.55–1.35 (br, 12H), 1.00–0.95 (br, 6H).

^{13}C NMR (125 MHz, $C_2D_2Cl_4$, 120 °C): δ 140.6, 135.9, 135.0, 129.6, 126.7, 126.3, 126.1, 125.8, 31.4, 30.1, 29.4, 28.9, 22.3, 13.6. (Multiple signals were observed because of slow rotation of bonds.)

$M_n = 9.6$ kg/mol, $M_w = 19.5$ kg/mol, $M_w/M_n = 2.02$, DP = 23.

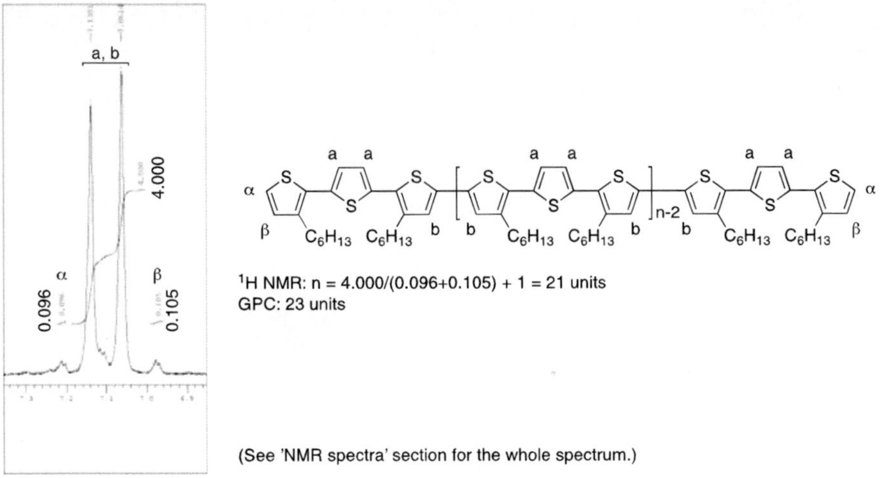

Supplementary Figure 4.1. Determination of DP of polymer **8** by ^{1}H NMR.

Polymer 9:

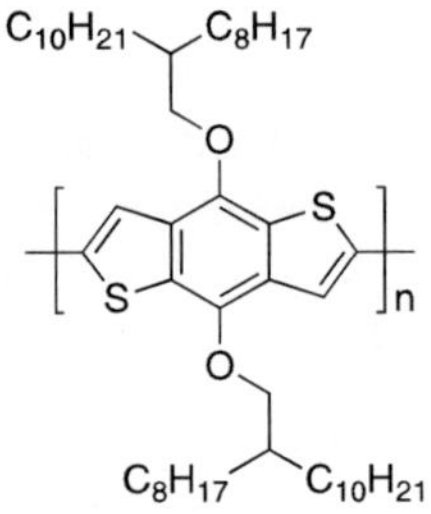

The title compound was obtained as dark red solid in 88% yield. The reaction was performed according to Procedure B. Small peaks observed in the ^{1}H NMR spectrum were assigned to the termini of the polymer. By integration of these peaks, DP was calculated to be 2.000/((0.074 + 0.076)/2) + 1 = 28.

^{1}H NMR (500 MHz, $C_2D_2Cl_4$, 120 °C): δ 7.73 (br s, 2H), 4.45–4.30 (br, 4H), 2.05–2.00 (br, 2H), 1.75–1.25 (br, 64H), 0.97–0.87 (br, 12H).

^{13}C NMR (125 MHz, $C_2D_2Cl_4$, 120 °C): δ 144.4, 136.7, 132.4, 129.3, 118.3, 76.8, 39.4, 31.6 (two signals overlapped), 31.5 (two signals overlapped), 29.9 (two signals overlapped), 29.4 (four signals overlapped), 29.0 (two signals overlapped), 26.9 (two signals overlapped), 22.3 (two signals overlapped), 13.6 (two signals overlapped).

M_n = 17.2 kg/mol, M_w = 37.0 kg/mol, M_w/M_n = 2.16, DP = 22.

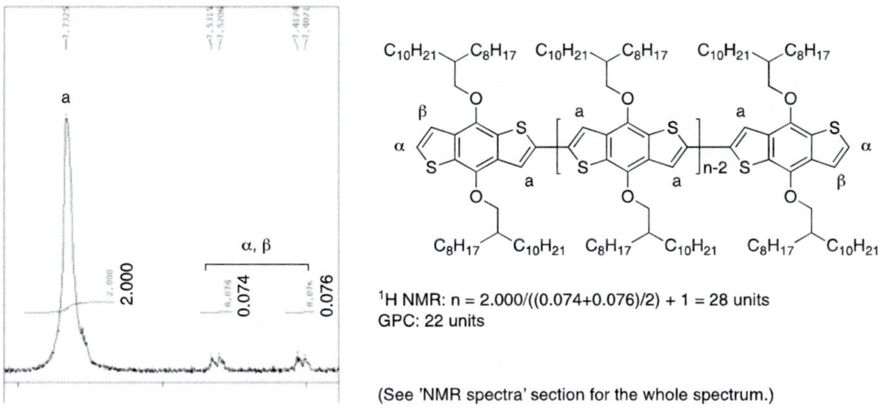

Supplementary Figure 4.2. Determination of DP of polymer **9** by ^{1}H NMR.

Polymer 10:

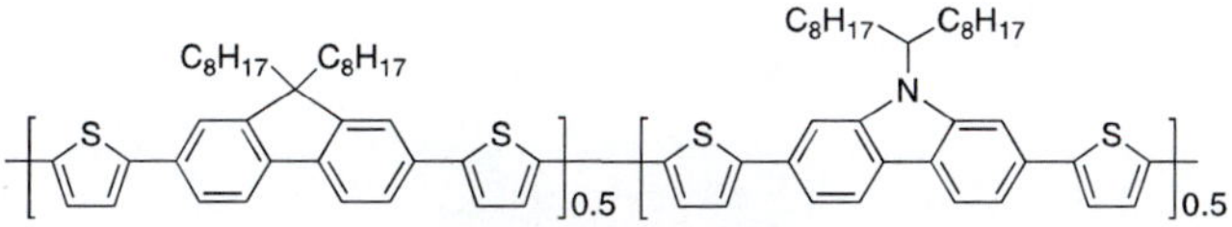

In an oven-dried Schlenk tube was added 2,2'-(9,9-dioctyl-9*H*-fluorene-2,7-diyl)dithiophene (55 mg, 0.10 mmol), 9-(heptadecan-9-yl)-2,7-di(thiophen-2-yl)-9*H*-carbazole (57 mg, 0.10 mmol) and benzofuryl-TP (20 mg, 0.030 mmol), and the Schlenk tube was evacuated and refilled with argon for three times. Then a THF solution of $FeCl_3{\bullet}6H_2O$ (0.067 mol/L, 0.30 mL, 0.020 mmol) was added and a toluene solution of $AlMe_3$ (2.0 mol/L, 0.30 mL, 0.60 mmol) was added by rinsing the wall of the Schlenk tube at rt. Finally, diethyl oxalate (54 μ L, 0.40 mmol) was added and the reaction mixture was degassed by Freeze–Pump–Thaw cycling for three times. After stirring with an agitation rate of 500 rpm at 70 °C for 24 h, the reaction mixture was diluted with THF (2 mL) and quenched by addition of a dioxane solution of HCl (4 mol/L, 0.50 mL), and stirred for 1 h. The crude solution was transferred to a 15 mL centrifuge tube and further diluted with THF up to 5 mL. Water was added up to 10 mL, and the tube was shaken vigorously after which orange polymer precipitated immediately. The mixture was centrifuged (4000 rpm, 10 min), and the solution was discarded. Then THF/water (1:1) was added up to 10 mL and the tube was shaken vigorously and centrifuged (4000 rpm, 10 min), and the solution was discarded. Finally, methanol was added up to 10 mL and the tube was shaken vigorously. The solid was filtered and dried under vacuum overnight to afford the product as orange solid (110 mg, 98%). A chloroform solution of the product was prepared and analyzed by GPC to determine the M_n, M_w, M_w/M_n values of the isolated polymer. ^{13}C NMR spectrum could not be analyzed because of the weak intensity of the signals.

^{1}H NMR (500 MHz, $C_2D_2Cl_4$, 120 °C): δ 8.10 (br d, $J = 8.0$ Hz, 2H), 7.78–7.72 (br, 4H), 7.68–7.63 (br, 4H), 7.55 (br d, $J = 8.0$ Hz, 2H), 7.40–7.33 (br, 4H), 7.32–7.25 (br, 4H), 4.70–4.65 (br, 1H), 2.45–2.40 (br, 2H), 2.15–2.05 (br, 6H), 1.40–1.15 (br, 44H), 0.95–0.85 (br, 16H).

$M_n = 19.7$ kg/mol, $M_w = 38.0$ kg/mol, $M_w/M_n = 1.93$, DP = 35.

Polymer 11:

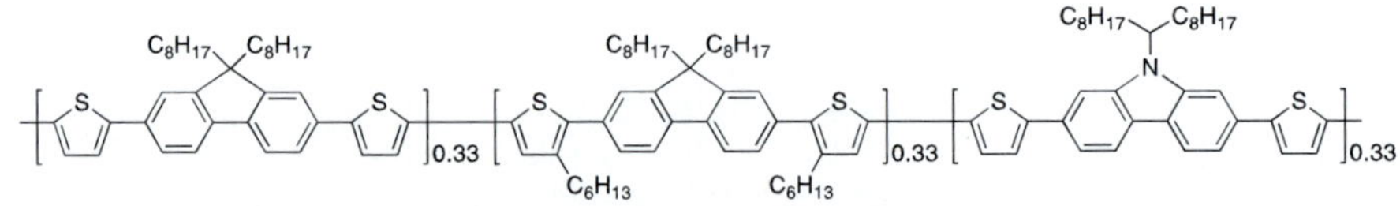

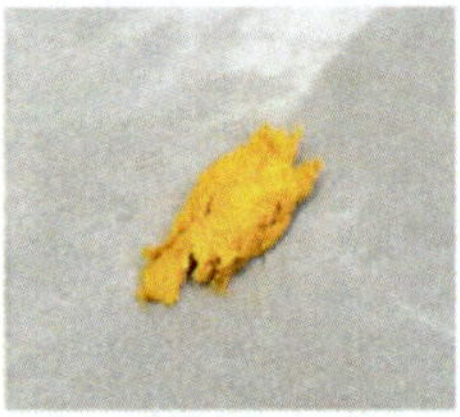

In an oven-dried Schlenk tube was added 2,2'-(9,9-dioctyl-9*H*-fluorene-2,7-diyl)dithiophene (37 mg, 0.067 mmol), 2,2'-(9,9-dioctyl-9*H*-fluorene-2,7-diyl)bis(3-hexylthiophene) (48 mg, 0.067 mmol), 9-(heptadecan-9-yl)-2,7-di(thiophen-2-yl)-9*H*-carbazole (38 mg, 0.10 mmol) and benzofuryl-TP (20 mg, 0.030 mmol), and the Schlenk tube was evacuated and refilled with argon for three times. Then a THF solution of $FeCl_3{\bullet}6H_2O$ (0.067 mol/L, 0.30 mL, 0.020 mmol) was added and a toluene solution of $AlMe_3$ (2.0 mol/L, 0.30 mL, 0.60 mmol) was added by rinsing the wall of the Schlenk tube at rt. Finally, diethyl oxalate (54 μ L, 0.40 mmol) was added and the reaction mixture was degassed by Freeze–Pump–Thaw cycling for three times. After stirring with an agitation rate of 500 rpm at 70 °C for 24 h, the reaction mixture was diluted with THF (2 mL) and quenched by addition of a dioxane solution of HCl (4 mol/L, 0.50 mL), and stirred for 1 h. The crude solution was diluted with chloroform (10 mL) and washed with water. The solution was filtered through a membrane filter to give a clear solution. The organic phase was dried over Na_2SO_4 and passed through a pad of Florisil. The solvent was removed under reduced pressure, and the crude product was purified by gel permeation chromatography (chloroform) to afford the product as yellow solid (68 mg, 56%). A chloroform solution of the product was prepared and analyzed by GPC to determine the M_n, M_w, M_w/M_n values of the isolated polymer. ^{13}C NMR spectrum could not be analyzed because of the weak intensity of the signals.

^1H NMR (500 MHz, $C_2D_2Cl_4$, 120 °C): δ 8.12–8.06 (br, 2H), 7.81–7.72 (br, 6H), 7.68–7.61 (br, 4H), 7.58–7.48 (br, 6H), 7.39–7.32 (br, 4H), 7.31–7.22 (br, 4H), 7.20–7.13 (br, 2H), 4.70–4.65 (br, 1H), 2.80–2.75 (br, 4H), 2.40–2.35 (br, 2H), 2.15–2.05 (br, 10H), 1.75–1.70 (br, 4H), 1.45–1.10 (br, 76H), 0.95–0.85 (br, 32H).

$M_n = 23.1$ kg/mol, $M_w = 56.3$ kg/mol, $M_w/M_n = 2.44$, DP = 38.

Procedure for control experiments (Table 4.4)

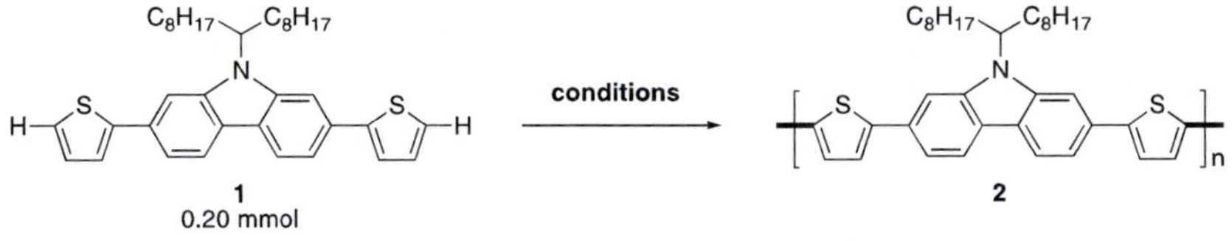

entry	conditions	M_n (kg/mol)[a]	M_w (kg/mol)[a]	M_w/M_n[a]	DP[b]
1	$FeCl_3\cdot 6H_2O$ (10 mol %), Benzofuryl-TP (15 mol %), $AlMe_3$ (3.0 equiv), $(COOEt)_2$ (2.0 equiv), THF (0.30 mL), PhMe (0.30 mL), 70 °C, 24 h	13.3	22.6	1.70	23
2	$Pd(OAc)_2$ (10 mol %), K_2CO_3 (2.1 equiv), $Cu(OAc)_2$ (2.0 equiv), DMA (2.5 mL), 110 °C, 72 h.	2.0	3.3	1.64	4
3	$Pd(OAc)_2$ (5.0 mol %), KOAc (2.0 equiv), Ag_2CO_3 (2.0 equiv), DMA (2.5 mL), 110 °C, 72 h.	3.7	9.2	2.53	7

[a] M_n and M_w were determined by GPC using polystyrene standards for the crude polymers.
[b] Degree of polymerization was calculated using M_n.

Procedure for entry 2[7b]: In an oven-dried Schlenk tube was added 9-(heptadecan-9-yl)-2,7-di(thiophen-2-yl)-9*H*-carbazole (114 mg, 0.20 mmol), $Cu(OAc)_2$ (73 mg, 0.40 mmol), and K_2CO_3 (58 mg, 0.42 mmol). DMA (2.0 mL) was added, and the mixture was stirred at 110 °C for 10 min, then $Pd(OAc)_2$ (4.5 mg, 0.020 mmol) dissolved in 0.50 mL of DMA was added to the reaction flask and reacted for 72 h. The mixture was cooled to rt, diluted with chloroform, and analyzed by GPC to determine the M_n, M_w, M_w/M_n values. Degree of polymerization (DP) was determined by dividing the M_n value by the molecular weight of the monomer unit.

Procedure for entry 3[7c]: In an oven-dried Schlenk tube was added a monomer (0.20 mmol), Ag_2CO_3 (110 mg, 0.40 mmol), KOAc (39 mg, 0.40 mmol), and DMA (2.0 mL). The mixture was stirred at 110 °C for 10 min, then $Pd(OAc)_2$ (2.2 mg, 0.010 mmol) dissolved in 0.50 mL of DMA was added to the reaction flask and reacted for 72 h. The mixture was cooled to rt, diluted with chloroform, and analyzed by GPC to determine the M_n, M_w, M_w/M_n values. Degree of polymerization (DP) was determined by dividing the M_n value by the molecular weight of the monomer unit.

Procedure for plotting M_n and M_w/M_n values as a function of monomer conversion (Fig. 4.6).

In an oven-dried Schlenk tube was added 2,2'-(9,9-dioctyl-9*H*-fluorene-2,7-diyl)dithiophene (111 mg, 0.20 mmol) and benzofuryl-TP (13 mg, 0.020 mmol), and the Schlenk tube was evacuated and refilled with argon for three times. Then a THF solution of $FeCl_3{\bullet}6H_2O$ (0.033 mol/L, 0.30 mL, 0.010 mmol) was added, and a toluene solution of $AlMe_3$ (2.0 mol/L, 0.30 mL, 0.60 mmol) was added by rinsing the wall of the Schlenk tube. Finally, diethyl oxalate (54 μ L, 0.40 mmol) was added and the reaction mixture was degassed by Freeze–Pump–Thaw cycling for three times. The reaction mixture was stirred in agitation rate of 500 rpm at 70 °C and sampled at the reaction time of 0.5, 1, 2, 4, 6, 8, 24, and 74 h. These samples were quenched by addition of a dioxane solution of HCl (4 mol/L, 0.10 mL), diluted with $CHCl_3$ (2 mL), sonicated for 15 min, washed with water, and analyzed by GPC using polystyrene standards.

Procedure for removal of residual catalyst from polymer (Fig. 4.8)

	residual element (ppm[a])	
	before treatment	after treatment
Fe	721	36
P	4502	175
Al	-[b]	-[b]

[a] Calculated based on [amount of residual element (mg)]/[amount of polymer (kg)].
[b] Determination of Al content in polymer was hindered because of the presence of Al in high concentration (~1.5 ppm) even in the blank sample. 1.5 ppm of Al in ICP sample corresponds to 1500 ppm of Al in 10 mg of polymer.

Polymer **3** (52 mg, M_n = 16.4 kg/mol, M_w = 36.3 kg/mol, M_w/M_n = 2.21, DP = 30) was dissolved in chloroform (20 mL) under sonication for 15 min and thiol-functionalized silica scavenger (1.0 g, Fuji Silysia Chemical Ltd.) was added. After stirring for 1 h, the solution was filtered and thiol-functionalized silica scavenger (1.0 g) was added. After stirring for another 1 h, the solution was filtered and thiol-functionalized silica scavenger (1.0 g) was added. After stirring for another 1 h, the solution was filtered and the amount of chloroform was reduced to 4 mL under reduced pressure. Methanol (20 mL) was added, and the solid was filtered and dried under vacuum overnight to afford the product as yellow solid (19 mg, 36% recovery, M_n = 18.7 kg/mol, M_w = 40.5 kg/mol, M_w/M_n = 2.17, DP = 34).

For ICP analysis, a polymer sample (~10 mg) was heated in sulfuric acid (0.20 mL) at 200 °C and nitric acid was added intermittently until a clear colorless solution was obtained. After evaporating the remaining nitic acid at 200 °C, the solution was cooled to rt and diluted to 10 mL with pure water. ICP analysis was conducted using standard solutions of Fe, P, and Al (blank, 0.10, 0.20, 0.50, 1.0 ppm for each element) for calibration.

When the same polymer **3** (26 mg) in chloroform (10 mL) was treated only once with thiol-functionalized silica scavenger (0.50 g), the polymer was recovered in 59% and the Fe content was 43 ppm.

For comparison, ICP analysis was conducted for the commercially available F8T2 (Sigma-Aldrich Co., LCC) and the Pd content was determined to be 222 ppm.

Without this post-treatment procedure, most of the other polymers contained over 1000 ppm of residual iron. For device applications, this procedure is recommended to ensure reproducibility.

Preparation of Mes-TADHT

***N*,*N*-bis(4-(3-hexylthiophen-2-yl)phenyl)-2,4,6-trimethylaniline**:

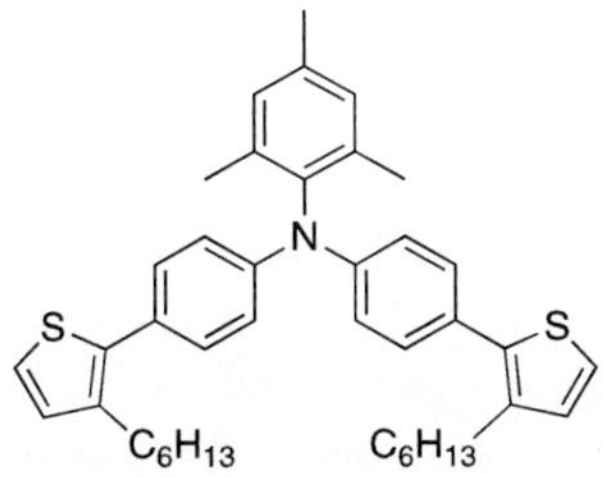

A mixture of *N*,*N*-bis(4-bromophenyl)-2,4,6-trimethylaniline (0.89 g, 2.0 mmol), 2-(3-hexylthiophen-2-yl)-4,4,5,5-tetramethyl-1,3,2-dioxaborolane (1.77 g, 6.0 mmol), K_2CO_3 (1.11 g, 8.0 mmol), $Pd_2(dba)_3$ (37 mg, 0.040 mmol) and SPhos (33 mg, 0.080 mmol) in degassed THF/H_2O (2:1, 4.5 mL) was stirred under argon atmosphere at 70 °C for 18 h. The reaction mixture was cooled to room temperature and extracted with ethyl acetate (10 mL × 3). The combined organic layers were dried over Na_2SO_4. The solvent was removed under reduced pressure, and the crude product was purified by silica gel chromatography (hexane only to hexane: dichloromethane = 19:1) to afford the product as colorless oil (1.16 g, 94%).

^{1}H NMR (500 MHz, $C_2D_2Cl_4$, 120 °C): δ 7.31 (m, 4H), 7.20–7.13 (br, 2H), 7.11–7.00 (br, 4H), 6.99 (s, 2H), 6.96 (d, $J = 5.2$ Hz, 2H), 2.70 (br, 4H), 2.37 (s, 3H), 2.09 (s, 6H), 1.66–1.60 (m, 4H), 1.38–1.31 (m, 12H), 0.95–0.89 (m, 6H). (Broad signals were observed because of slow rotation of bonds.)

^{13}C NMR (125 MHz, $C_2D_2Cl_4$, 120 °C): δ 145.2, 138.0, 137.9, 137.4, 136.7, 130.1, 129.6, 127.4, 120.2, 119.9, 119.6, 119.1, 118.9, 31.3, 30.5, 28.8, 28.6, 22.2, 20.7, 18.1, 13.6. (Multiple signals were observed because of slow rotation of bonds.)

HRMS (APCI +): *m/z* calcd for $C_{41}H_{49}NS_2$ [M + H$^+$] 620.3379; found: 620.3369.

Mes-TADHT:

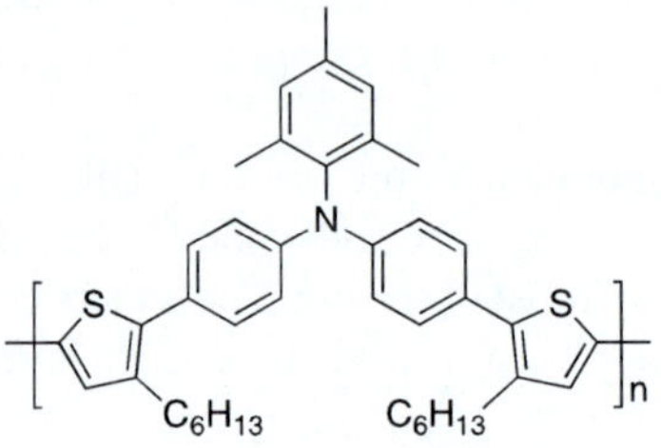

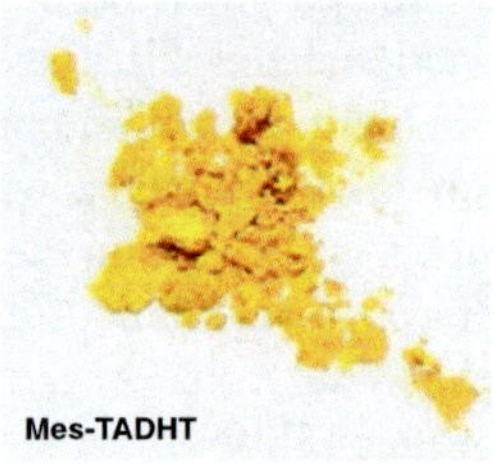

In an oven-dried Schlenk tube was added *N*,*N*-bis(4-(3-hexylthiophen-2-yl)phenyl)-2,4,6-trimethylaniline (248 mg, 0.40 mmol) and benzofuryl-TP (40 mg, 0.060 mmol), and the Schlenk tube was evacuated and refilled with argon for three times. Then a THF solution of $FeCl_3$•$6H_2O$ (0.067 mol/L, 0.60 mL, 0.040 mmol) was added, and a toluene solution of $AlMe_3$ (2.0 mol/L, 0.60 mL, 1.2 mmol) was added by rinsing the wall of the Schlenk tube at rt. Finally, diethyl oxalate (108 mL, 0.80 mmol) was added and the reaction mixture was degassed by Freeze–Pump–Thaw cycling for three times. After stirring with an agitation rate of 500 rpm at 70 °C for 24 h, the reaction mixture was diluted with THF (2 mL) and quenched by addition of a dioxane solution of HCl (4 mol/L, 1.0 mL), and stirred for 1 h. The crude solution was diluted with chloroform (20 mL) and washed with water (20 mL). The solution was filtered through a membrane filter to give a clear solution. The organic phase was dried over Na_2SO_4 and passed through a pad of Florisil. The solvent was removed under reduced pressure, and the crude product was purified by gel permeation chromatography (chloroform). The chloroform solution of polymer was added a thiol-functionalized silica scavenger (2.0 g, Fuji Silysia Chemical Ltd.), and the mixture was stirred at rt for 1 h and filtered. This post-treatment procedure was repeated again, and most of the solvent was removed under reduced pressure. Finally, methanol (50 mL) was added and the precipitate was filtered and dried under vacuum overnight to afford the product as yellow solid (114 mg, 46% yield).

^{1}H NMR (500 MHz, $C_2D_2Cl_4$, 25 °C): δ 7.29–7.27 (br, 4H), 7.03–7.01 (br, 4H), 7.00 (br s, 2H), 6.96 (br s, 2H), 2.65–2.60 (br, 4H), 2.33 (br s, 3H), 2.03 (br s, 6H), 1.65–1.55 (br, 4H), 1.35–1.20 (br, 12H), 0.85 (br t, $J = 6.9$ Hz, 6H).

^{13}C NMR (125 MHz, $C_2D_2Cl_4$, 25 °C): δ 144.8, 139.4, 138.7, 137.4, 136.4, 134.3, 129.9, 129.5, 126.6, 125.8, 120.2, 119.4, 31.5, 30.8, 29.1, 28.8, 22.5, 21.1, 18.5, 14.1.

$M_n = 13.0$ kg/mol, $M_w = 26.4$ kg/mol, $M_w/M_n = 2.03$, DP = 21.

Materials and methods for device fabrication and evaluation

Note: All of the device fabrication and evaluation were conducted by Dr. Hao-Sheng Lin in Matsuo Lab. I deeply thank him for his generous help.

Unless otherwise noted, all materials including dry solvents were obtained from commercial suppliers (Adamas-beta, TCI) and used without further purification. C_{60} and $PC_{61}BM$ were purchased from American dye Inc. PbI_2 (99.9985%), and methylammonium iodide (MAI) was purchased from Sigma-Aldrich Inc. Anhydrous *N,N*-dimethylformamide (DMF), dimethyl sulfoxide (DMSO), *ortho*-dichlorobenzene (*o*-DCB), and chlorobenzene (CB) were purchased from Alfa Aesar. Poly(triaryl amine) (PTAA) was purchased from Merck.

SnO_2 precursor solution preparation: 27.1 mg $SnCl_2 \bullet 2H_2O$ (Aldrich, >99.995%) white powder was dissolved in 4.0 mL of anhydrous ethanol (TCI), which was filtered by 0.45 mm syringe filter before using.

$MAPbI_3$ precursor solution preparation: 355 mg of PbI_2, 122 mg of CH_3NH_3I, and 54.7 mL of DMSO (molar ratio 1:1:1) were mixed in 490.5 mL of DMF solution at room temperature with stirring for 1 h. The solution was filtered through a 0.45 mm polytetrafluoroethylene filter prior to use.

HTM solution preparation. A solution was prepared by mixing 10.0 mg HTM, different mass ratio of a stock solution of 170 mg mL^{-1} lithium bis(trifluoromethylsulfonyl)-imide (LiTFSI) in anhydrous acetonitrile (5.9 mL for 10 wt%; 11.8 mL for 20 wt%; 17.7 mL for 30 wt%), and 7.0 mL of 4-*tert*-butylpyridine in 1.0 mL anhydrous chlorobenzene.

The current density vs voltage (*J–V*) characteristics were measured using a software-controlled source meter (Keithley 2400 SourceMeter) under dark conditions and the simulated sunlight irradiation of 1 sun (AM 1.5G; 100 mW cm^{-2}) using a solar simulator (EMS-35AAA, Ushio Spax Inc.) with an Ushio Xe short arc lamp 500. The source meter was calibrated using a silicon diode (BS-520BK, Bunkokeiki). When evaluating, devices were masked with a black aperture to set the active area of the device to 0.09 cm^2. The scan directions of forward (from –0.2 to 1.2 V) and reverse (from 1.2 to –0.2 V) were used. The forward scan was measured followed by a reverse scan. A scan rate of 100 mV/s with a 0.02 step and 100 ms waiting time was used for the measurement. The incident photon-to-current conversion efficiency (IPCE) spectra were measured using machine spectrometer with a wavelength ranging from 300 to 850 nm.

Procedure for device fabrication

Indium-doped tin oxide (ITO) patterned glass substrates were cleaned and sonicated with detergent, distilled water, acetone, and isopropanol in an ultrasonic bath for 15 min, respectively. Next, the cleaned ITO substrates were treated with UV/O_3 for 15 min. Subsequently, 25 mL of SnO_2 precursor solution was spin-coated on the cleaned ITO substrate at 3000 rpm for 30 s, which was annealed at 150 °C for 45 min. After cooling down to room temperature, the spin-coating process was repeated one more time followed by annealing at 190 °C for 1 h. Then, the SnO_2-coated ITO glass was further treated with UV/O_3 for 15 min before spin coating of fullerene solution. Next, 25 mL of perovskite precursor solution was spin-coated on the fullerene layer at 4000 rpm for 30 s, with slowly dropping 0.5 mL of anhydrous diethyl ether onto the substrate 10 s after the start of the spin-coating process, followed up with annealing at 65 °C for 1 min and then 100 °C for 5 min. The hole-transporting layer was spin-coated from the 25 mL of prepared HTM solution at 3000 rpm for 30 s. Finally, a 70-nm-thick of Au anode was fabricated by thermal deposition at a constant evaporation rate of 0.05 nm s^{-1} under pressure of 10^{-6} Torr.

NMR spectra

^{1}H NMR spectra of polymer **2** and its monomer

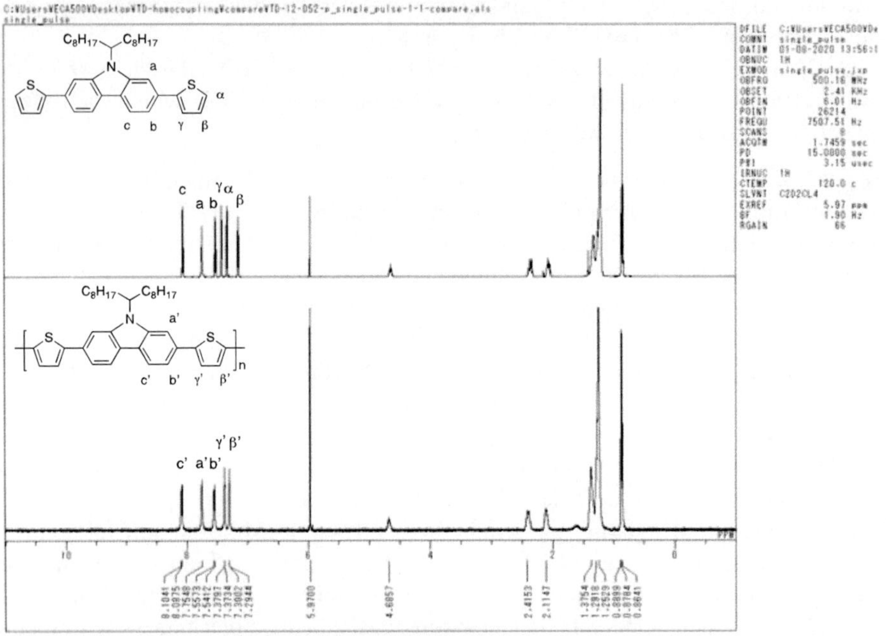

^{1}H NMR spectra of polymer **3** and its monomer

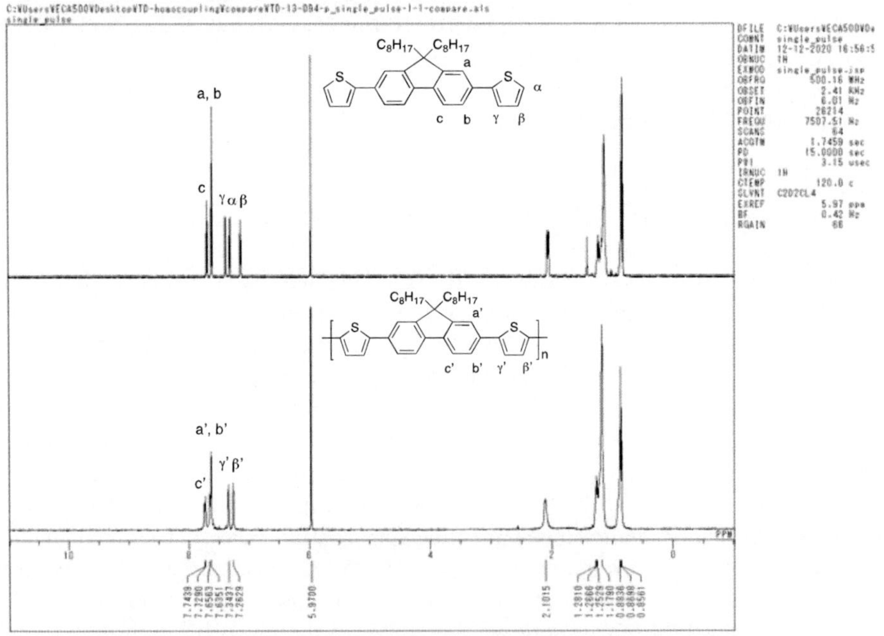

^{1}H NMR spectra of polymer **4** and its monomer

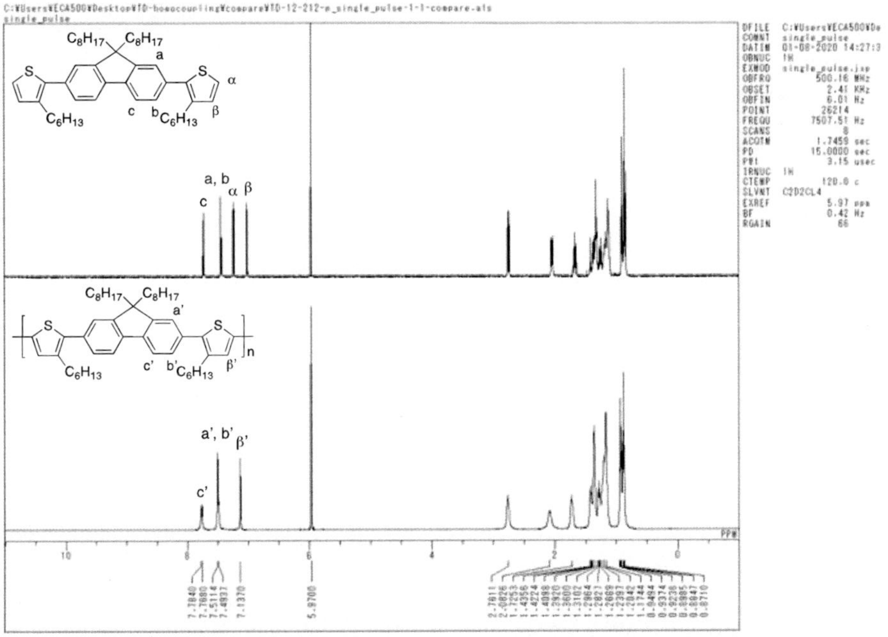

^{1}H NMR spectra of polymer **5** and its monomer

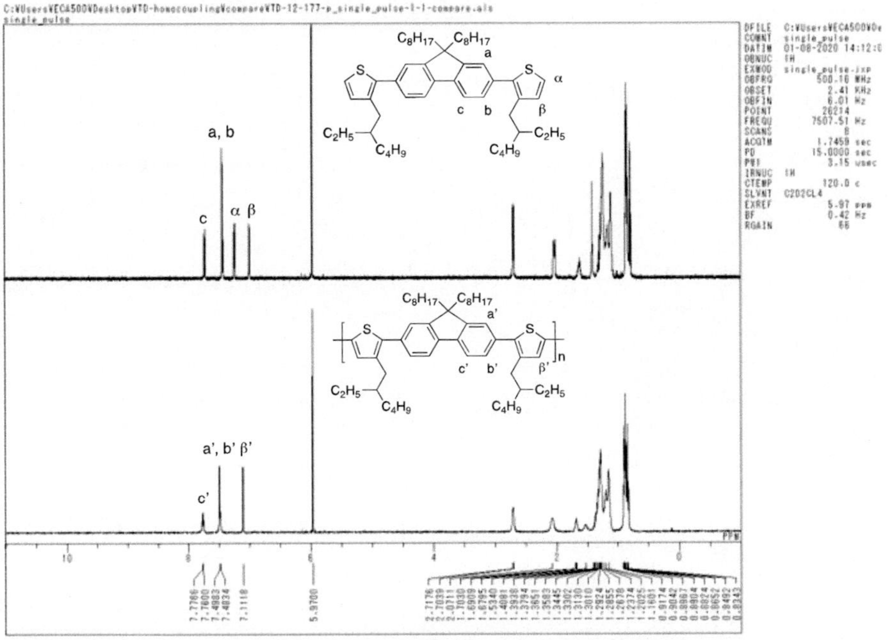

^{1}H NMR spectra of polymer **6** and its monomer

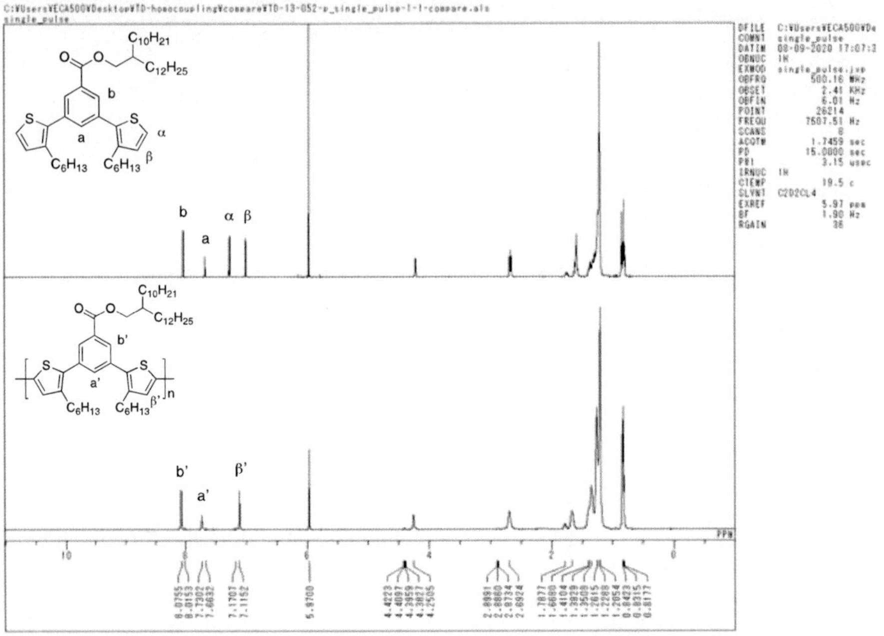

^{1}H NMR spectra of polymer **7** and its monomer

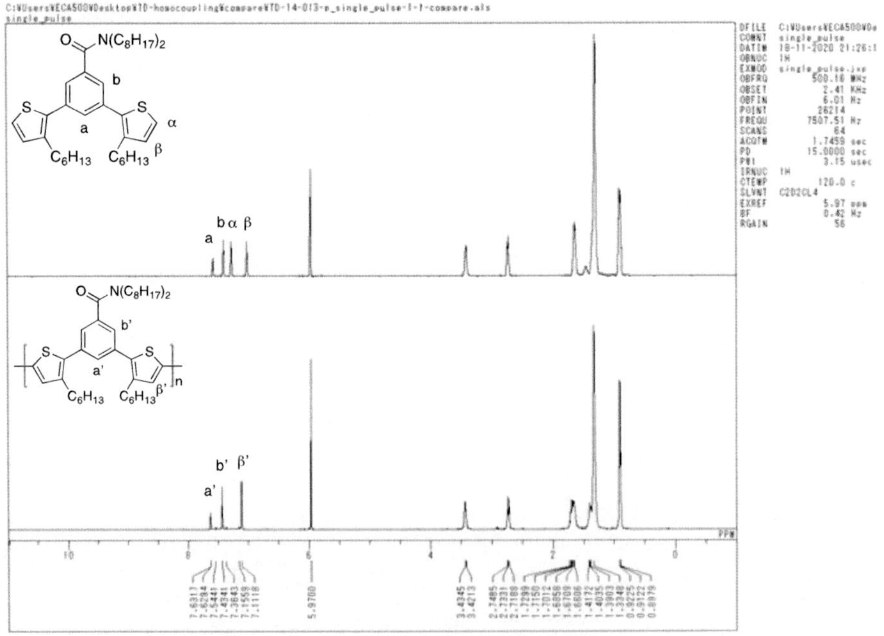

^{1}H NMR spectra of polymer **8** and its monomer

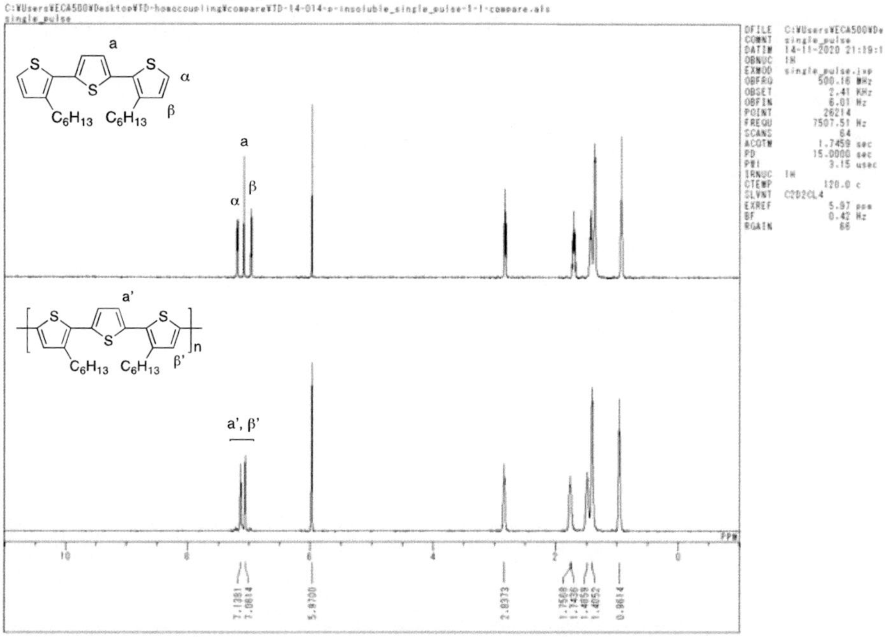

^{1}H NMR spectra of polymer **9** and its monomer

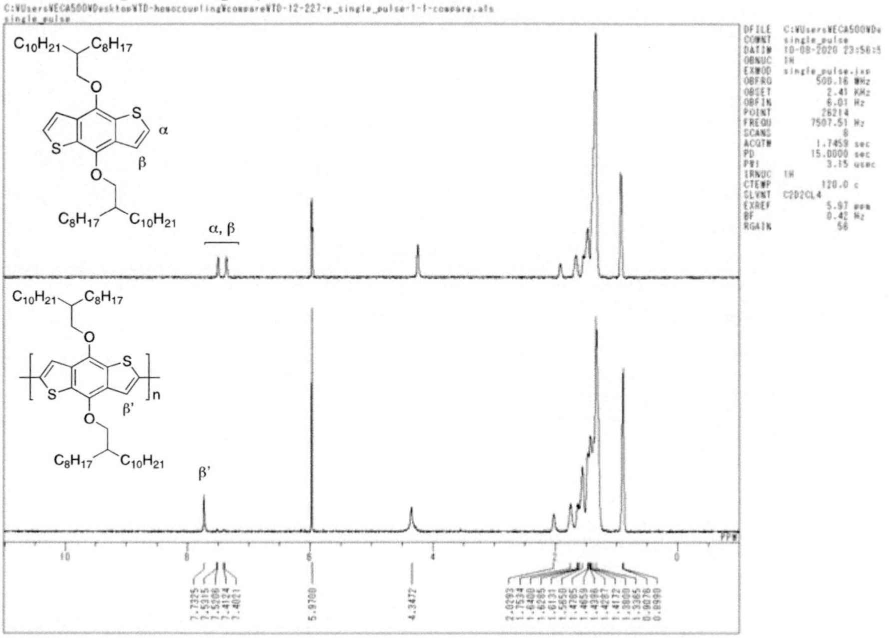

^{1}H NMR spectra of copolymer **10**, polymer **2**, and polymer **3**

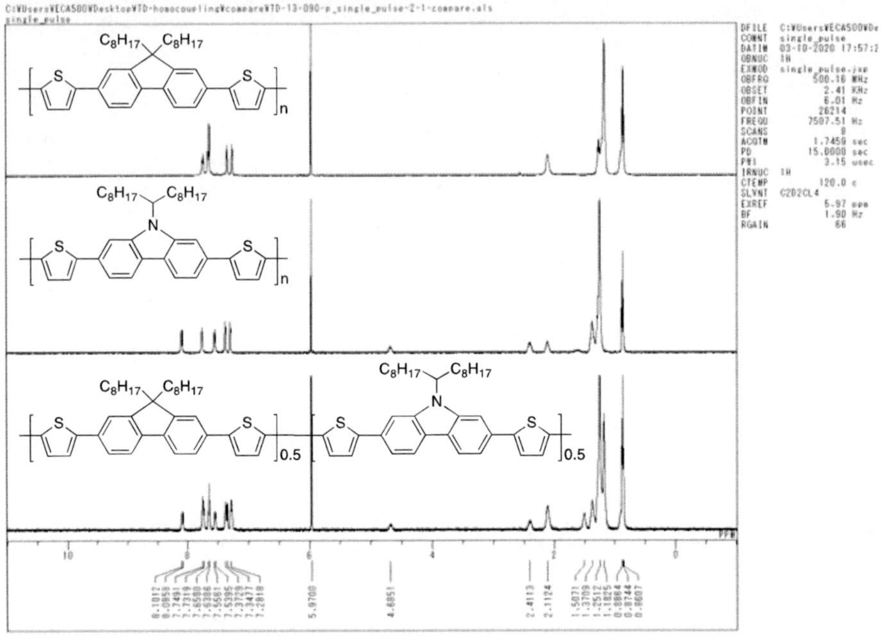

^{1}H NMR spectra of copolymer **11**, polymer **2**, polymer **3**, and polymer **4**

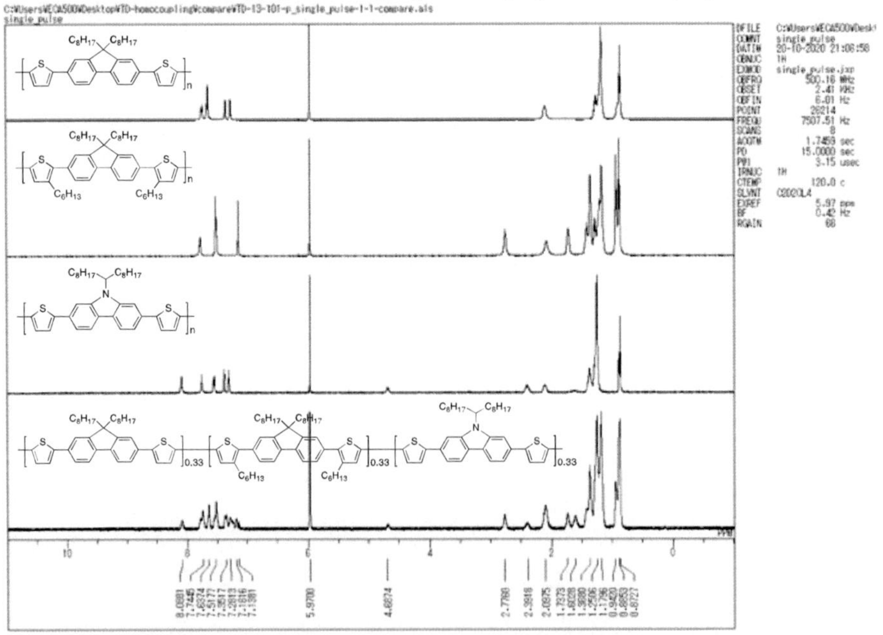

^{1}H NMR spectra of polymer **Mes-TADHT** and its monomer

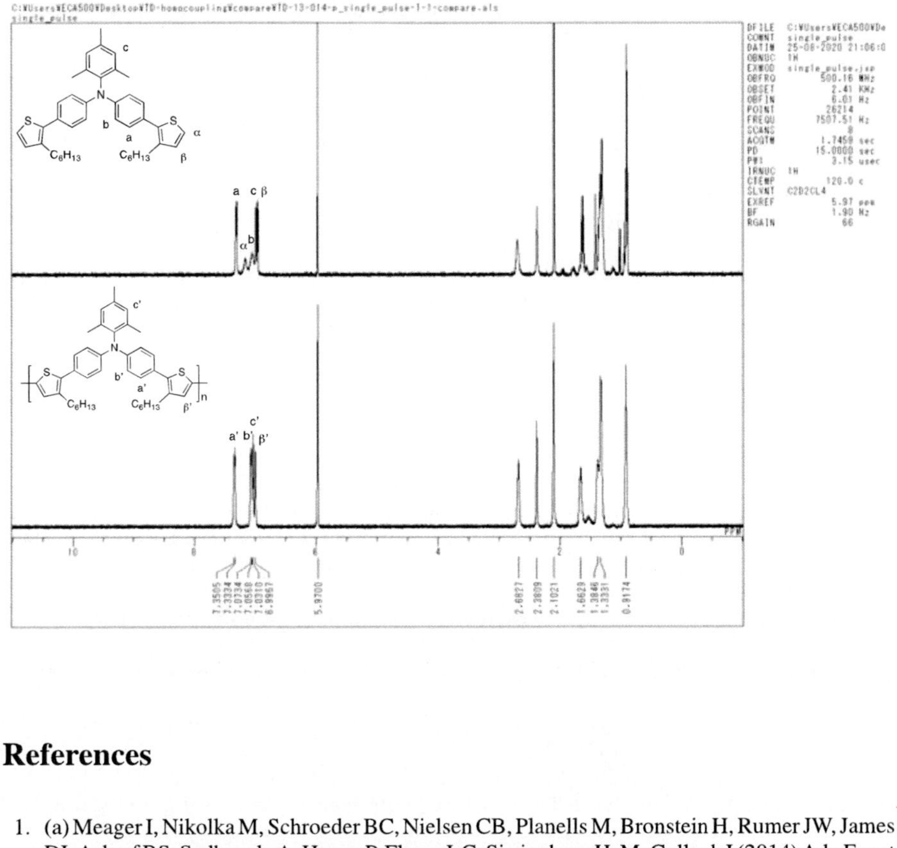

References

1. (a) Meager I, Nikolka M, Schroeder BC, Nielsen CB, Planells M, Bronstein H, Rumer JW, James DI, Ashraf RS, Sadhanala A, Hayoz P, Flores J-C, Sirringhaus H, McCulloch I (2014) Adv Funct Mater 24:7109–7115 (b) Durban MM, Kazarinoff PD, Luscombe CK (2010) Macromolecules 43:6348–6352 (c) Usta H, Lu G, Facchetti A, Marks TJ (2006) J Am Chem Soc 128:9034–9035 (d) Holliday S, Donaghey JE, McCulloch I (2014) Chem Mater 26:647–663 (e) Guo X, Baumgarten M, Müllen K (2013) Progress Poly Sci 38:1832–1908 (f) Huang H, Yang L, Facchetti A, Marks TJ, Chem Rev 117:10291–10318 (g) Zhao X, Zhan X (2011) Chem Soc Rev 40:3728–3743 (h) Wu W, Liu Y, Zhu D (2010) Chem Soc Rev 39:1489–1502 (i) Rasmussen SC, Evenson SJ, McCausland CB (2015) Chem Commun 51:4528–4543 (j) Sirringhaus H (2014) Adv. Mater 26:1319–1335 (k) Sekine C, Tsubata Y, Yamada T, Kitano M, Doi S (2014) Sci Tech Adv Mat 15:034203 (l) AlSalhi MS, Alam J, Dass LA, Raja M (2011) Int J Mol Sci 12:2036–2054 (m) Boudreault P-LT, Najari A, Leclerc M (2011) Chem Mater 23:456–469 (n) Arias AC, MacKenzie JD, McCulloch I, Rivnay J, Salleo A (2010) Chem Rev 110:3–24 (o) Brabec CJ, Gowrisanker S, Halls JJM, Laird D, Jia S, Williams SP (2010) Adv Mater 22:3839–3856 (p) Helgesen M, Søndergaard R, Krebs FC (2010) J Mater Chem 20:36–60 (q) Li G, Zhu R, Yang Y (2012) Nat Photon 6:153–161 (r) Cheng Y-J, Yang S-H, Hsu C-S (2009) Chem Rev 109:5868–5923 (s) You J, Dou L, Hong Z, Li G, Yang Y (2013) Progress Poly Sci 38:1909–1928 (t) Ong BS, Wu Y, Li Y, Liu P, Pan H (2008) Chem Eur J 14:4766–4778 (u) Zhang Z, Wang J (2012) J Mater Chem 22:4178
2. (a) Carsten B, He F, Son HJ, Xu T, Yu L (2011) Chem Rev 111:1493–1528 (b) Sakamoto J, Rehahn M, Wegner G, Schlüter AD (2009) Macromol Rapid Commun 30:653–687 (c) Xu S, Kim EH, Wei A, Negishi E (2014) Sci Tech Adv Mat 15:044201

3. (a) Wang Q, Zhang B, Liu L, Chen Y, Qu Y, Zhang X, Yang J, Xie Z, Geng Y, Wang L, Wang F (2012) J Phys Chem C 116:21727–21733 (b) Park JK, Jo J, Seo JH, Moon JS, Park YD, Lee K, Heeger AJ, Bazan GC (2011) Adv Mater 23:2430–2435 (c) Kim Y, Cook S, Kirkpatrick J, Nelson J, Durrant JR, Bradley DDC, Giles M, Heeney M, Hamilton R, McCulloch I (2007) J Phys Chem C 111:8137–8141 (d) Kim JS, Lee Y, Lee JH, Park JH, Kim JK, Cho K (2010) Adv Mater 22:1355–1360 (e) Mao Z, Vakhshouri K, Jaye C, Fischer DA, Fernando R, DeLongchamp DM, Gomez ED, Sauvé G (2013) Macromolecules 46:103–112 (f) Koldemir U, Puniredd SR, Wagner M, Tongay S, McCarley TD, Kamenov GD, Müllen K, Pisula W, Reynolds JR (2015) Macromolecules 48:6369–6377
4. Yang Y, Nishiura M, Wang H, Hou Z (2018) Coord Chem Rev 376:506–532
5. (a) Pouliot J-R, Grenier F, Blaskovits JT, Beaupré S, Leclerc M (2016) Chem Rev 116:14225–14274 (b) Bura T, Blaskovits JT, Leclerc M (2016) J Am Chem Soc 138:10056–10071 (c) Rudenko AE, Thompson BC (2015) J Polym Sci Part A: Polym Chem 53:135–147 (d) Dudnik AS, Aldrich TJ, Eastham ND, Chang RPH, Facchetti A, Marks TJ (2016) J Am Chem Soc 138:15699–15709 (e) Bura T, Morin P-O, Leclerc M (2015) Macromolecules 48:5614–5620 (f) Mercier LG, Leclerc M (2013) Acc Chem Res 46:1597–1605 (g) Matsidik R, Komber H, Luzio A, Caironi M, Sommer M (2014) J Am Chem Soc 137:6705–6711 (h) Wakioka M, Ozawa F (2018) Asian J Org Chem 7:1206–1216 (i) Kuwabara J, Yasuda T, Choi SJ, Lu W, Yamazaki K, Kagaya S, Han L, Kanbara T (2014) Adv Funct Mater 24:3226–3233
6. Tsuchiya K, Ogino K (2013) Polym J 45:281–286
7. (a) Gobalasingham NS, Noh S, Thompson BC (2016) Polym Chem 7:1623–1631 (b) Zhang Q, Wan X, Lu Y, Li Y, Li Y, Li C, Wu H, Chen Y (2014) Chem Commun 50:12497–12499 (c) Zhang Q, Li Y, Lu Y, Zhang H, Li M, Yang Y, Wang J, Chen Y, Li C (2015) Polymer 68:227–233
8. Huang Q, Qin X, Li B, Lan J, Guo Q, You J (2014) Chem Commun 50:13739–13741
9. Guo Q, Jiang R, Wu D, You J (2016) Macromol Rapid Commun 37:794–798
10. Wakim S, Blouin N, Gingras E, Tao Y, Leclerc M (2007) Macromol Rapid Commun 28:1798–1803
11. Ziebart C, Federsel C, Anbarasan P, Jackstell R, Baumann W, Spannenberg A, Beller M (2012) J Am Chem Soc 134:20701–20704
12. Ikariya T, Yamamoto AJ (1976) Organometal Chem 118:65–77
13. Donat-Bouillud A, Lévesque I, Tao Y, D'Iorio M, Beaupré S, Blondin P, Ranger M, Bouchard J, Leclerc M (2000) Chem Mater 12:1931–1936
14. Roncali J (1992) Chem Rev 92:711–738
15. (a) Carothers WH (1931) Chem Rev 8:353–426 (b) Flory PJ (1946) Chem Rev 39:137–197
16. (a) Sheina EE, Liu J, Iovu MC, Laird DW, McCullough RD (2004) Macromolecules 37:3526–3528 (b) Yokoyama A, Miyakoshi R, Yokozawa T (2004) Macromolecules 37:1169–1171 (c) Miyakoshi R, Yokoyama A, Yokozawa T (2004) Macromol Rapid Commun 25:1663–1666 (d) Miyakoshi R, Yokoyama A, Yokozawa T (2005) J Am Chem Soc 127:17542–17547 (e) Iovu MC, Sheina EE, Gil RR, McCullough RD (2005) Macromolecules 38:8649–8656 (f) Yokoyama A, Suzuki H, Kubota Y, Ohuchi K, Higashimura H, Yokozawa T (2007) J Am Chem Soc 129:7236–7237
17. (a) Urien M, Wantz G, Cloutet E, Hirsch L, Tardy P, Vignau L, Cramail H, Parneix J-P (2007) Org Electron 8:727–734 (b) Camaioni N, Tinti F, Franco L, Fabris M, Toffoletti A, Ruzzi M, Montanari L, Bonoldi L, Pellegrino A, Calabrese A, Po R (2012) Org Electron 13:550–559 (c) Cowan SR, Leong WL, Banerji N, Dennler G, Heeger A (2011) J Adv Funct Mater 21:3083–3092 (d) Nikiforov MP, Lai B, Chen W, Chen S, Schaller RD, Strzalka J, Maser J, Darling SB (2013) Energy Environ Sci 6:1513–1520 (e) Ashraf RS, Schroeder BC, Bronstein HA, Huang Z, Thomas S, Kline RJ, Brabec CJ, Rannou P, Anthopoulos TD, Durrant JR, McCulloch I (2013) Adv Mater 25:2029–2034 (f) Estrada LA, Deininger JJ, Kamenov GD, Reynolds JR (2013) ACS Macro Lett 2:869–873 (g) Usluer Ö, Abbas M, Wantz G, Vignau L, Hirsch L, Grana E, Brochon C, Cloutet E, Hadziioannou G (2014) ACS Macro Lett 3:1134–1138
18. Still WC, Kahn M, Mitra A (1978) J Org Chem 43:2923–2925

19. Pangborn AB, Giardello MA, Grubbs RH, Rosen RK, Timmers FJ (1996) Organometallics 15:1518–1520
20. Doba T, Ilies L, Sato W, Shang R, Nakamura E (2021) Nat Catal 4:631–638
21. Gallazzi MC, Castellani L, Zerbi G, Sozzani P (1991) Synth Met 41:495–498

Chapter 5
Iron-Catalyzed Oxidative C–H Alkenylation of Thiophenes with Enamines

5.1 Introduction

Transition-metal-catalyzed oxidative C–H alkenylation has attracted much attention as one of the most straightforward methods to synthesize alkenylated compounds. In 1967, Fujiwara and Moritani reported that styrene/$PdCl_2$ complex reacts with a solvent amount of benzene to afford *trans*-stilbene in low yield (Scheme 5.1) [1].

Two years later, they reported that AgOAc can be used as an oxidant to achieve palladium-catalyzed oxidative C–H alkenylation of a solvent amount of benzene with a limiting amount of styrene (Scheme 5.2) [2]. Therefore, nowadays oxidative C–H alkenylation reaction is called as Fujiwara–Moritani reaction. A proposed reaction mechanism is shown in Fig. 5.1. First, $Pd(OAc)_2$ undergoes C–H activation of an arene through concerted metalation-deprotonation (CMD) mechanism [3] to generate an aryl palladium species (**II**). Then an alkene inserts into the C–Pd bond to give an alkyl palladium species (**III**). **III** undergoes β-H elimination to afford the alkenylated product and a palladium hydride (**IV**). Reductive elimination and reoxidation of **V** to **I** by AgOAc close the catalytic cycle.

Soon after directing group strategy [4] appeared, de Vries and van Leeuwen reported *ortho*-selective oxidative C–H alkenylation using anilide as a directing group (Scheme 5.3) [5]. In this reaction, only 1 equiv of a C–H substrate is required because of the directing group effect. After this preliminary result, *ortho*-selective oxidative C–H alkenylation using a directing group has been extensively studied [6].

In 2012, Yu and coworkers reported *meta*-selective oxidative C–H alkenylation assisted by an end-on template (Scheme 5.4a) [7, 8]. Based on this report, Maiti and coworkers extended this idea to *para*-selective oxidative C–H alkenylation (Scheme 5.4b) [9].

Although directing group strategy achieved a great success in this field, non-directed oxidative C–H alkenylation faced a limitation of requiring a solvent amount of the C–H substrate. In 2017, Yu and coworkers succeeded in reducing the amount of the C–H substrate to a limiting amount, i.e., to 1 equiv, using pyridone as a ligand

T. Doba, *Iron-Catalyzed C–H/C–H Coupling for Synthesis of Functional Small Molecules and Polymers*, Springer Theses,
https://doi.org/10.1007/978-981-99-4121-6_5

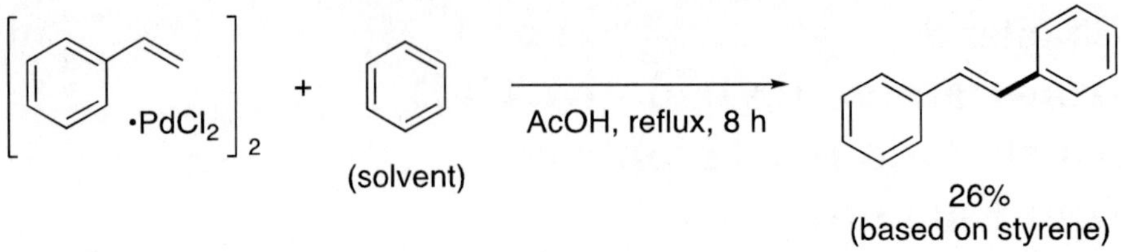

Scheme 5.1 Palladium-mediated oxidative C–H alkenylation

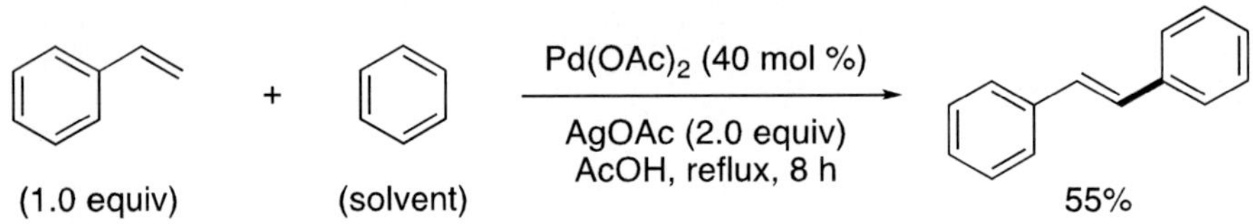

Scheme 5.2 Palladium-catalyzed oxidative C–H alkenylation

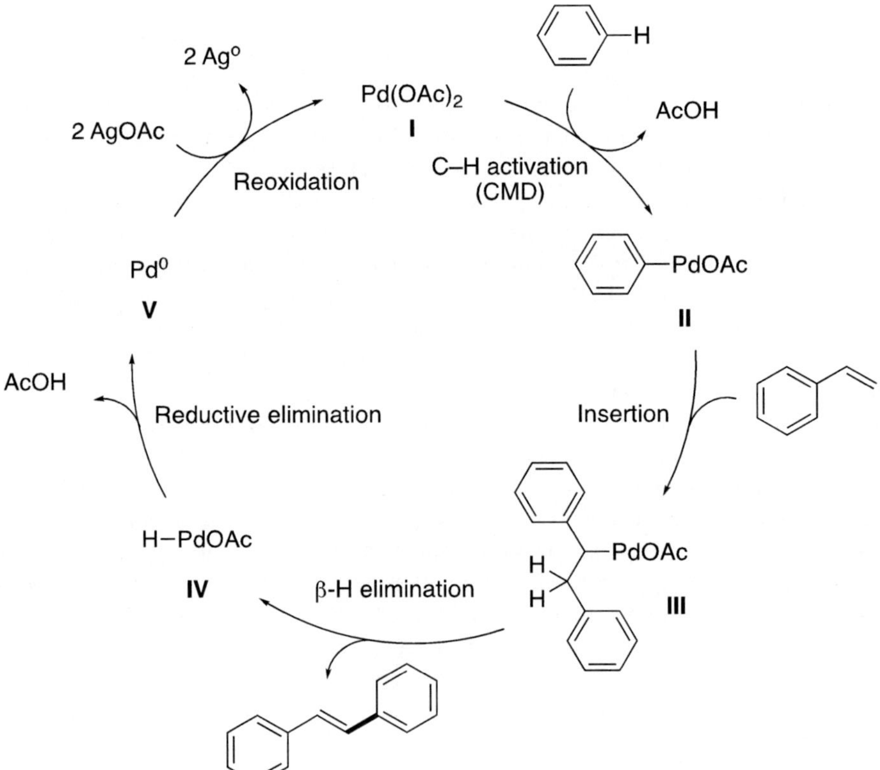

Fig. 5.1 A proposed mechanism

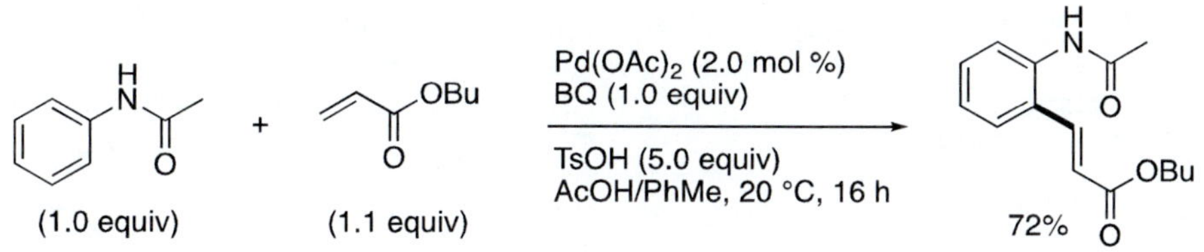

Scheme 5.3 Palladium-catalyzed ortho-selective oxidative C–H alkenylation

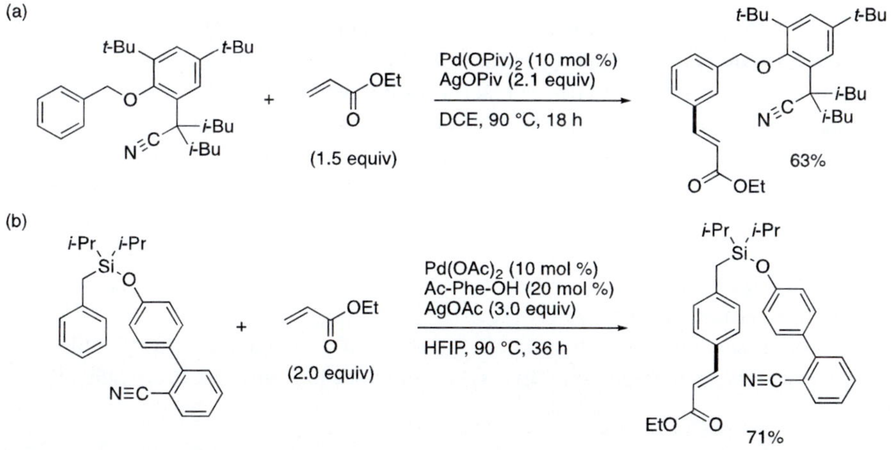

Scheme 5.4 Palladium-catalyzed meta- and para-selective oxidative C–H alkenylation

(Scheme 5.5) [10, 11]. The site selectivity was governed by a combination of steric and electronic effects.

As described above, since the initial discovery of transition-metal-catalyzed oxidative C–H alkenylation by Fujiwara and Moritani, various reaction systems have been developed to expand the scope of arenes. However, the scope of alkenes employed has long been limited to electron-deficient alkenes such as acrylates and styrenes, and there are only a limited number of reports on the reaction with electron-rich ones. The difficulty of using electron-rich alkenes is ascribed to a high insertion barrier of organometallic species to a negatively charge alkene in the carbometallation step (Scheme 5.6) [12].

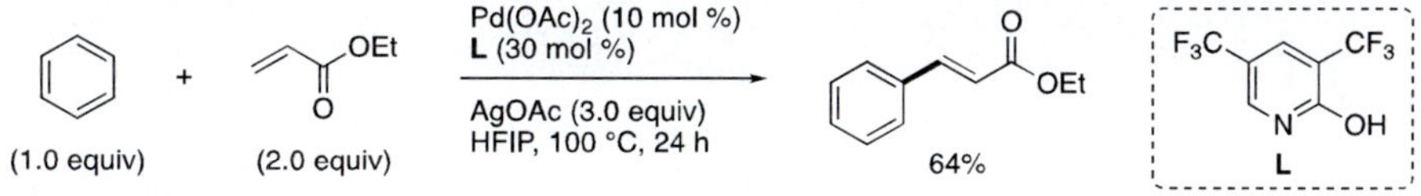

Scheme 5.5 Palladium-catalyzed oxidative C–H alkenylation with a limiting amount of arene

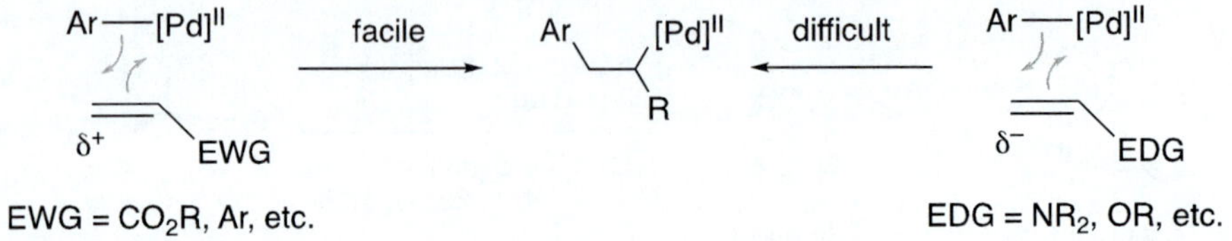

Scheme 5.6 Carbometallation of organometallic species to electron-deficient and electron-rich alkenes

In 2010, Bergman and Ellman reported rhodium-catalyzed oxidative C–H alkenylation with unactivated alkenes (Scheme 5.7). In this reaction, d^6-cationic Rh(III) was chosen as a catalyst to facilitate the insertion step by weakening the π-back donation and decreasing the electron density of a coordinating alkene. After this work, several groups have reported the reaction with unactivated alkenes [13].

Incorporation of electron-rich alkenes has been recognized as a challenging issue, and there are only a limited number of reports on the reaction with electron-rich alkenes [14]. In 2018, Zhang and coworkers reported rhodium-catalyzed oxidative C–H alkenylation with electron-rich alkenes such as vinyl ethers and an *N*-vinylphthalimide (Scheme 5.8) [15].

Although several reactions with electron-rich alkenes have been reported so far, transition-metal-catalyzed oxidative C–H alkenylation with electron-rich alkenes

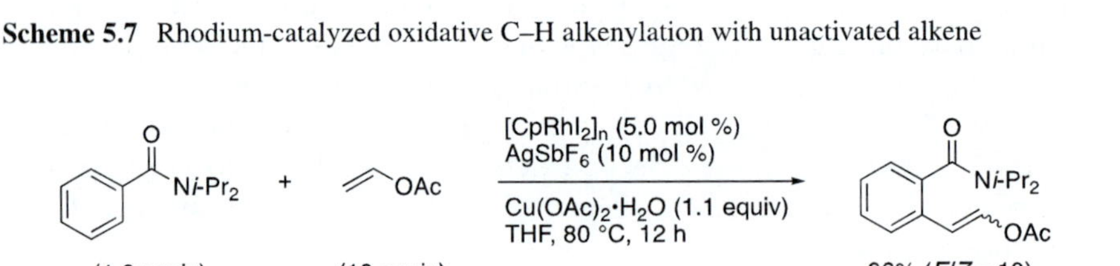

Scheme 5.7 Rhodium-catalyzed oxidative C–H alkenylation with unactivated alkene

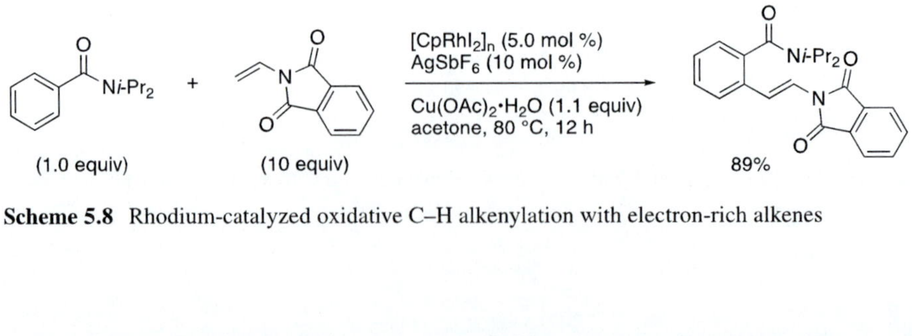

Scheme 5.8 Rhodium-catalyzed oxidative C–H alkenylation with electron-rich alkenes

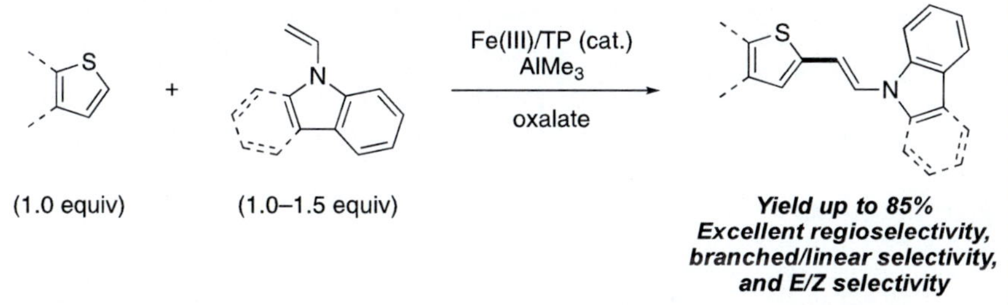

Scheme 5.9 Iron-catalyzed oxidative C–H alkenylation of thiophenes with vinylcarbazoles and vinylindoles

under controlled regioselectivity, branched/linear selectivity, and *E*/*Z* selectivity is still recognized as a formidable challenge. In this chapter, I report iron-catalyzed oxidative C–H alkenylation of thiophenes with vinylcarbazoles and vinylindoles, which is an important class of electron-rich moieties in materials science (Scheme 5.9) [16]. This reaction shows excellent regioselectivity, branched/linear selectivity, and *E*/*Z* selectivity without the aid of a preinstalled directing group.

5.2 Reaction Design and Initial Results

Based on the proposed reaction mechanism of iron-catalyzed regioselective thienyl C–H/C–H homocoupling, I speculated that the electron-deficient iron(III) intermediate formed after the C–H activation of a thiophene (**VI**) is suitable for the subsequent reaction with electron-rich alkenes through an electrophilic metalation (**VI** to **VII**)/deprotonation (**VII** to **VIII**) mechanism [17] to give the oxidative C–H alkenylation product (Fig. 5.2). High LUMO energy of electron-rich alkene also helps to prevent the catalyst poisoning described in Chap. 2 by weakening the π-back donation from a low-valent iron species generated after reductive elimination.

Based on this hypothesis, I investigated the reaction of 2-phenylthiophene with various kinds of electron-rich alkenes using 10 mol % of $Fe(acac)_3$ as a catalyst, 11 mol % of TP as a ligand, 3.0 equiv of $AlMe_3$ as a base, and 2.0 equiv of $(COOEt)_2$ as an oxidant in a mixed solvent of THF/PhMe at 70 °C for 15 h. As shown in Table 5.1, the desired oxidative C–H alkenylation proceeded in high yield with 9-vinylcarbazole, accompanying the formation of homocoupling products of both 2-phenylthiophene and 9-vinylcarbazole (**4**, **5**). This reaction showed excellent regioselectivity, branched/linear selectivity, and *E*/*Z* selectivity. Simple *N*-phenyl-*N*-vinylaniline (**6**) was not as reactive as 9-vinylcarbazole because of the instability of the enamine and the product under the reaction conditions. 1,1-disubstituted alkenes (**7**, **8**), 1,2-disubstituted alkenes (**3**, **9**), and a 1,1,2-trisubstituted alkene (**10**) were all unreactive possibly due to the steric hindrance. Vinyl amides (**11**, **12**),

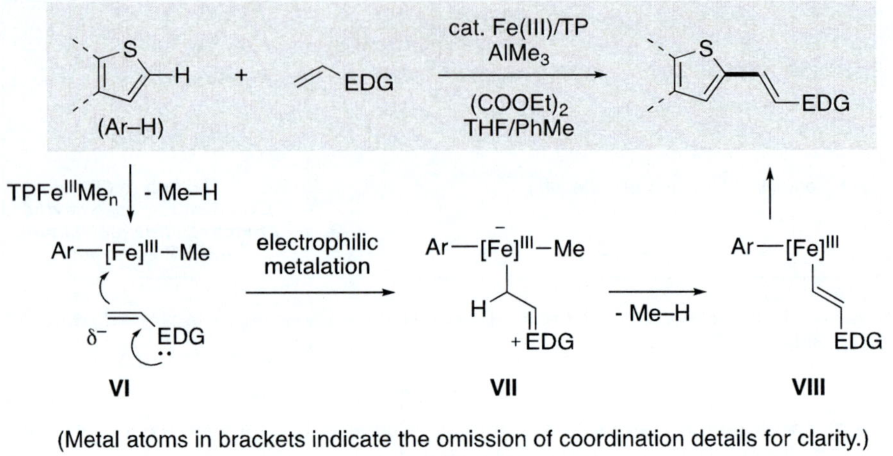

Fig. 5.2 A working hypothesis of iron-catalyzed oxidative C–H alkenylation of thiophenes with electron-rich alkenes

N-vinylphthalimide (**13**), acetaldehyde (**14**), and enol ethers (**15**, **16**) were all unreactive probably because of insufficient nucleophilicity. Other heteroatom-substituted alkenes (**17–20**), an unactivated alkene (**21**), styrene (**22**), and methyl acrylate (**23**) were not only unreactive but also inhibited the homocoupling of **1**. This may be because of the strong interaction of these alkenes with the low-valent iron species generated after reductive elimination.

Control experiments proved that all of the components are necessary for the reaction to take place (Table 5.2). Notably, THF was essential for this reaction, suggesting that coordinating ethereal solvent is required to stabilize the organoiron intermediates.

5.3 Reaction Kinetics

To probe when the byproducts are formed during the reaction, I investigated the reaction kinetics (Fig. 5.3). In region **A**, or at the beginning of the reaction, both of the homocoupling byproducts (**4**, **5**) are formed at similar rates but at lower rates compared to that of cross-coupling. In region **B**, or at the end of the reaction, once the thiophene is completely consumed, the remaining enamine gradually undergoes homocoupling. These results indicate that all of the catalytic cycles that generate the cross-coupling product (**3**) and the homocoupling products (**4**, **5**) are operative from the beginning to the end of the reaction without control.

Table 5.1 Initial results of iron-catalyzed oxidative C–H alkenylation of 2-phenylthiophene with enamines

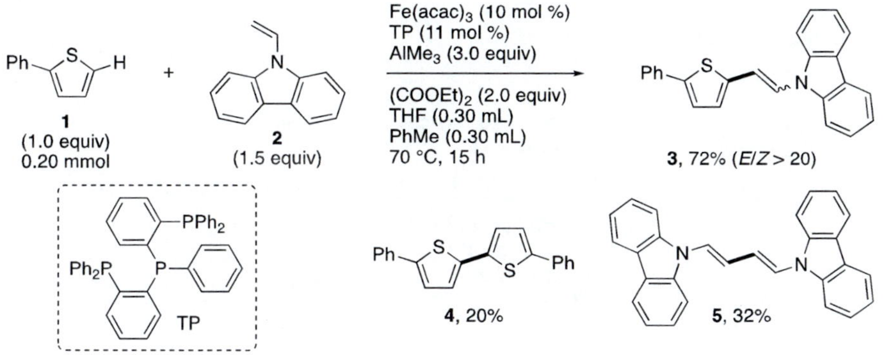

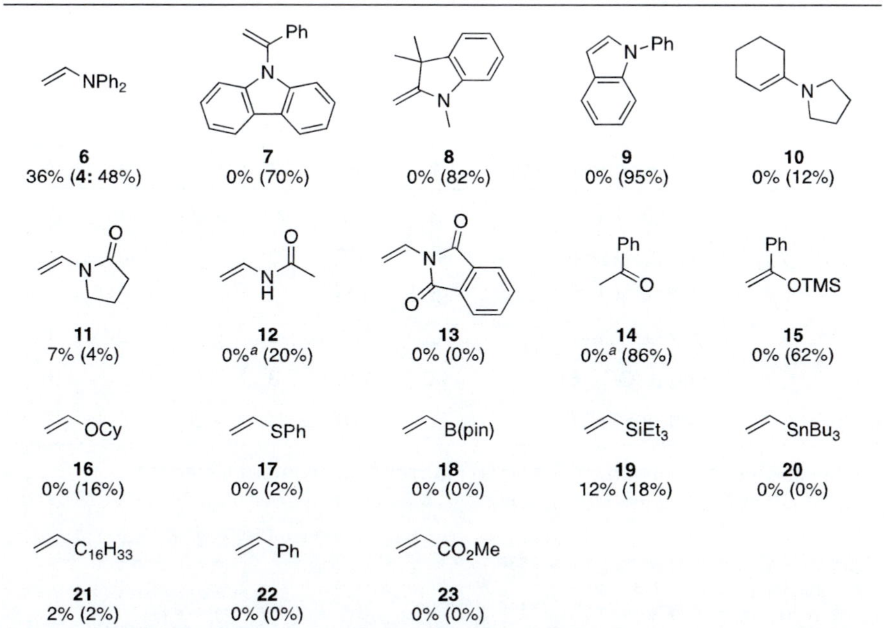

Yields were estimated by GC using tridecane as an internal standard. Yields of **4** were determined either by GC or ^{1}H NMR and are shown in parentheses. [a] 4.5 equiv of $AlMe_3$ was used.

5.4 Investigation of Reaction Parameters

To improve the homo/cross-coupling selectivity, I first investigated the effect of each reaction parameters using TP as a ligand. The easiest solution to affect the product distribution is to change the substrate ratio. By simply increasing the amount of **2**, homocoupling of **1** was suppressed (Table 5.3, entry 2, 3). However, conversion of **1** dropped because of the catalyst inhibition by alkene coordination. A slight excess of **1** can also be employed to afford the cross-coupling product in high yield with

Table 5.2 Control experiments

standard:

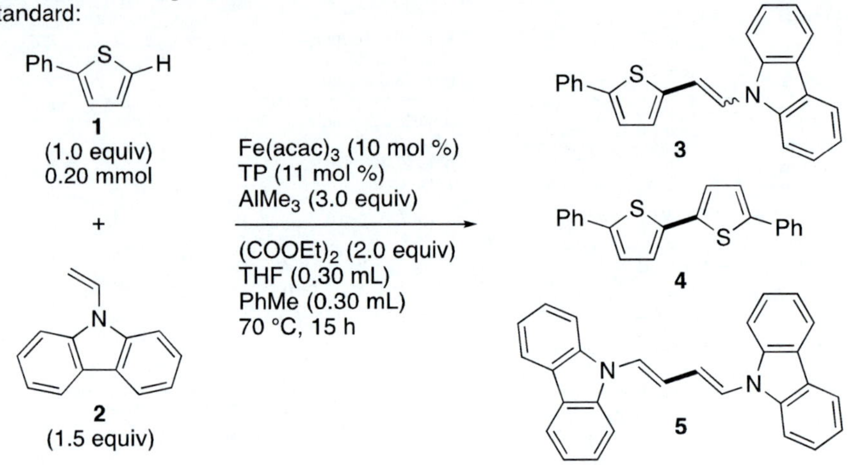

entry	variation of conditions	yield (%)[a] 1	2	3	*E*/*Z*	4	5
1	standard	3	30	72	> 20	20	32
2	without $Fe(acac)_3$	99	129	0	–	0	0
3	without TP	99	131	0	–	0	0
4	without $AlMe_3$	100	150	0	–	0	0
5	without $(COOEt)_2$	90	145	3	–	2	0
6	without THF[b]	65	119	19	> 20	8	0

[a] Yields were determined by GC using tridecane as an internal standard.
[b] 0.60 mL of PhMe was used in total.

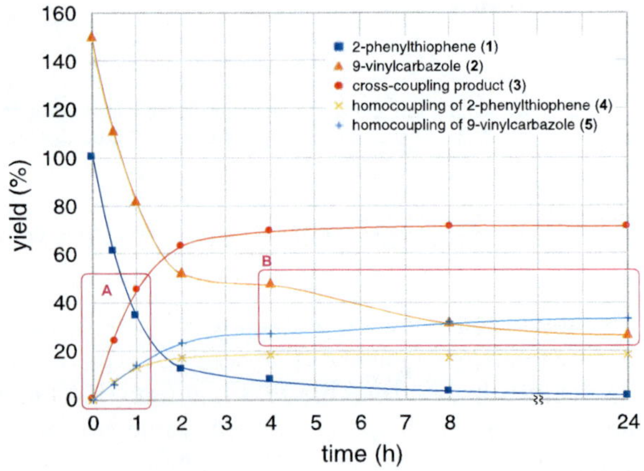

Fig. 5.3 Reaction kinetics. Reaction conditions: **1** (1.0 equiv, 0.20 mmol), **2** (1.5 equiv), $Fe(acac)_3$ (10 mol %), TP (11 mol %), $AlMe_3$ (3.0 equiv), $(COOEt)_2$ (2.0 equiv), THF (0.30 mL), PhMe (0.30 mL), 70 °C. Sampled at 0.5, 1, 2, 4, 8, 24 h

nearly an equal amount of thiophene homocoupling (entry 4). To let the reaction favor the reaction pathway with lower energy barrier and improve the homo/cross-coupling selectivity, reaction temperature was lowered. As expected, at 50 °C, the homo/cross-coupling selectivity was improved and the product was obtained in 84% yield with higher conversion of **1** and **2** (entry 6). This indicates that lower reaction temperature helps to suppress catalyst deactivation. 40 °C was not enough to achieve efficient catalyst turnover (entry 7). Variation of all other parameters was not effective for improving the homo/cross-coupling selectivity (entry 8–22). 3.0 mol % of catalyst was not enough to fully consume the starting material (entry 8). In contrast to the iron-catalyzed thienyl C–H/C–H polycondensation described in the previous chapter, iron chlorides were not effective for this reaction (entry 9, 10). 1.0 equiv of $AlMe_3$ was not enough, which suggests that only one methyl group from $AlMe_3$ is effectively transferred to iron and acts as a base (entry 11). The exact amount of $AlMe_3$ (2.0 equiv) did not decrease the yield (entry 12). Interestingly, $AlMe_2Cl$ was totally ineffective and resulted in decomposition of enamine possibly due to its higher Lewis acidity (entry 13). Other aluminum bases such as $AlEt_3$ and DIBAL-H were also ineffective (entry 14, 15), which can be explained by the formation of catalytically inactive iron hydride species. The exact amount of $(COOEt)_2$ (1.0 equiv) gave comparable results with the standard conditions with slightly higher recovery of **1** (entry 16). $(COOEt)_2$ was determined to be the best oxidant for this reaction (entry 17–19). Diluted conditions resulted in insufficient conversion (entry 20) and the solvent ratio had little effect on the product distribution (entry 21, 22).

5.5 Effect of Ligand

Based on the observation that iron-catalyzed regioselective thienyl C–H/C–H polycondensation proceeds less efficiently with more electron-rich TP ligands (Chap. 4), I investigated the effect of ligand on the homo/cross-coupling selectivity, assuming that more electron-rich TP ligands would suppress homocoupling of thiophene and give higher yields of the cross-coupling product. Unexpectedly, no significant difference of homo/cross-coupling selectivity was observed among Me_2N-TP, MeO-TP, TP, and F-TP (Table 5.4, entry 1–4). The use of F_3C-TP (entry 5) resulted in a slightly higher yield of **4**, which is in line with the observation that more electron-deficient TP ligands gave higher DP values in iron-catalyzed regioselective thienyl C–H/C–H polycondensation. Indolyl-TP resulted in insufficient conversion and benzofuryl-TP and benzothienyl-TP failed to improve the yield (entry 6–8). In summary, ligand structure had little effect on the homo/cross-coupling selectivity, indicating that the homo/cross-coupling selectivity is mainly governed by the reactivity of the substrates themselves.

Table 5.3 Effect of reaction parameters

standard:

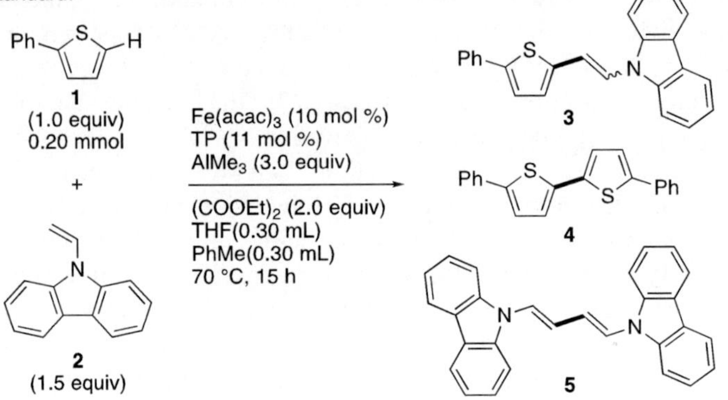

entry	variation of conditions	yield (%)[a] 1	2	3	*E*/*Z*	4	5
1	standard	3	30	72	> 20	20	32
2	**2** (1.0 equiv)	1	8	68	> 20	30	14
3	**2** (2.0 equiv)	10	80	71	> 20	15	32
4	**1** (1.5 equiv), **2** (1.0 equiv)	1	1	79	> 20	75[b]	10
5	60 °C	1	26	78	> 20	16	34
6	50 °C	0	15	84	> 20	14	46
7	40 °C	32	60	51	> 20	11	22
8	$Fe(acac)_3$ (3.0 mol %), TP (3.3 mol %)	52	78	31	17	12	8
9	$FeCl_3 \cdot 6H_2O$ instead of $Fe(acac)_3$	28	69	43	14	13	14
10	$FeCl_2 \cdot 4H_2O$ instead of $Fe(acac)_3$	36	72	42	17	15	13
11	$AlMe_3$ (1.0 equiv)	18	65	57	> 20	17	16
12	$AlMe_3$ (2.0 equiv)	2	36	72	> 20	18	28
13	AMe_2Cl instead of $AlMe_3$	96	4	0	–	0	0
14	$AlEt_3$ instead of $AlMe_3$	96	109	0	–	0	0
15	DIBAL-H instead of $AlMe_3$	95	119	0	–	0	0
16	$(COOEt)_2$ (1.0 equiv)	6	44	70	> 20	18	26
17	$(COOMe)_2$ instead of $(COOEt)_2$	21	49	55	> 20	18	20
18	$(COO\textit{t}\text{-}Bu)_2$ instead of $(COOEt)_2$	11	4	29	12	33	4
19	DCP instead of $(COOEt)_2$	48	79	36	> 20	11	8
20	THF (0.60 mL), PhMe (0.60 mL)	10	42	67	> 20	21	29
21	THF (0.60 mL)	8	38	70	> 20	21	29
22	PhMe (0.60 mL)	4	31	70	> 20	21	33

[a] Yields were determined by GC using tridecane as an internal standard.
[b] The yield was determined by ^1H NMR using 1,3,5-trimethoxybenzene as an internal standard.
DCP: 1,2-dichloropropane

Table 5.4 Effect of ligand on product selectivity

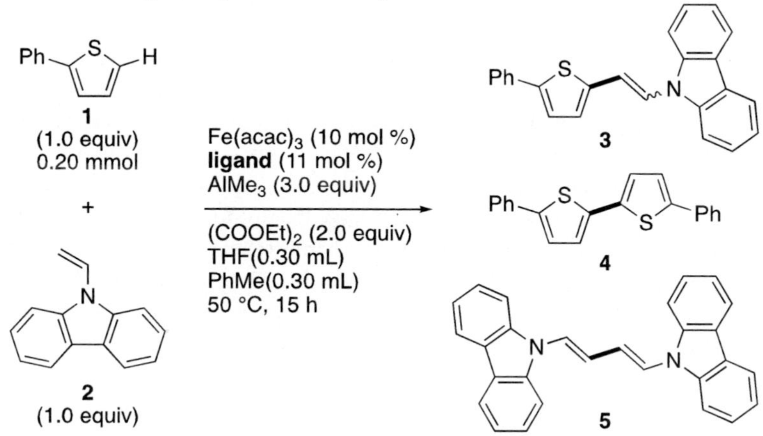

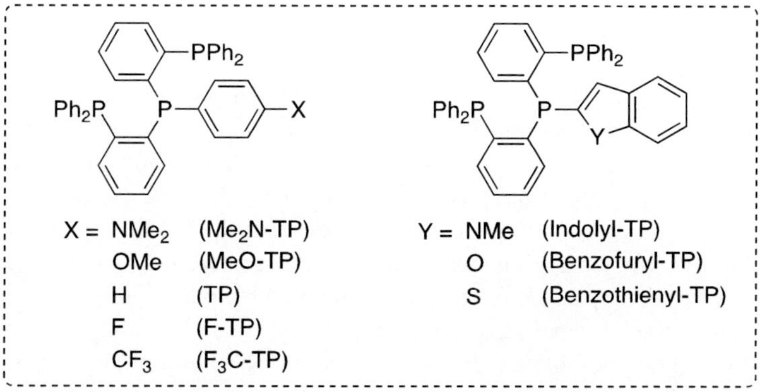

		yield (%)[a]					
entry	**ligand**	**1**	**2**	**3**	*E*/*Z*	**4**	**5**
1	Me_2N-TP	20	20	56	15	22	15
2	MeO-TP	5	10	68	> 20	25	17
3	TP	1	0	73	> 20	24	21
4	F-TP	22	28	54	> 20	21	13
5	F_3C-TP	4	20	66	> 20	30	11
6	Indolyl-TP	20	30	42	> 20	33	12
7	Benzofuryl-TP	10	22	61	> 20	26	6
8	Benzothienyl-TP	3	16	69	> 20	27	9

[a] Yields were determined by GC using tridecane as an internal standard.

5.6 Substrate Scope

With optimized reaction conditions in hand, the scope of iron-catalyzed oxidative C–H alkenylation of thiophenes with enamines was investigated (Table 5.5). Homo-coupling byproduct **4** or their analogues accounted for the mass balance in all cases. 2-phenylthiophene possessing either electron donating or withdrawing groups reacted well to afford the alkenylated product in high yields (**3**, **24**–**27**) with high to excellent *E* selectivity. 2-Hexylthiophene also reacted to give the product in high yield with excellent *E* selectivity (**28**). 2,3-Disubstituted thiophenes reacted under the same reaction conditions to afford the product in high yields with slightly lower *E* selectivity (**29**, **30**). Thiophenes possessing not only phenyl and alkyl but also heteroaryl groups such as benzofuryl and thienyl groups were also amenable (**31**, **32**). Triisopropylsilyl (**32**) and pinacoboronate (**33**) on arenes useful for cross-coupling reactions were well-tolerated. Notably, the EDOT core was alkenylated in high yield (**34**), which is in contrast with the iron-catalyzed regioselective thienyl C–H/C–H homocoupling. The very mild oxidative conditions in the presence of diethyl oxalate as an oxidant enabled the synthesis of oxidant-susceptible compounds containing carbazole (**35**) and electron-rich triarylamine (**36**, **37**) groups which potentially serve as a hole-transporting moiety. Compound **36** showed intense fluorescence at blue-green region with a quantum yield of $\Phi_{FL} = 0.84$, demonstrating the emissive properties of the obtained products. A substrate possessing tetraphenylbenzene moiety that cyclizes easily under oxidative conditions underwent selective thienyl C–H coupling in high yield (**38**), indicating orthogonal reactivity to Scholl reaction. 2,7- and 3,6-diarylated *N*-vinylcarbazoles reacted to afford highly conjugated vinylene products (**39**, **40**). Not only *N*-vinylcarbazoles but also *N*-vinylindoles (**41**, **42**), *N*-vinylpyrrole (**43**), and *N*-vinyldiphenylamine (**44**) were amenable, giving another option for this reaction.

5.7 Copolymerization

To demonstrate the applicability of this method for the synthesis of polymeric electronic materials, copolymers and an end-capped thiophene polymer were synthesized by C–H/C–H cross-coupling. Table 5.6 illustrates examples of copolymers obtained by straightforward reaction between a bisthiophene and a bisenamine monomers. GPC analyzes of these polymers referenced to polystyrene standards showed degree of polymerization (DP) up to 24 and a unimodal distribution with M_w/M_n values close to 2, indicating that the reaction takes place though a step-growth mechanism. Obtained polymers contained two monomer units in 1:1 ratio reflecting the initial reactant ratio. Reaction of monomers with an **F8T2** and a 9*H*,9'*H*-3,3'-bicarbazole (BCZ) structure, both of which are widely used in optoelectronic materials, smoothly gave the product in high yield (**45**). Mes-TABT structure, which has been recently reported to be effective for perovskite solar cells was also amenable

Table 5.5 Substrate scope of iron-catalyzed oxidative C–H alkenylation

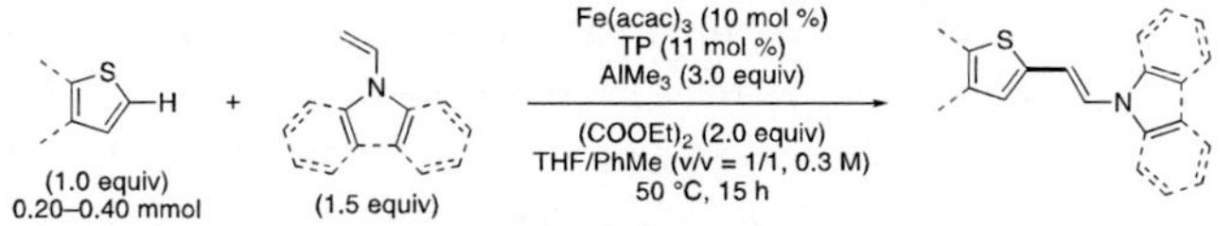

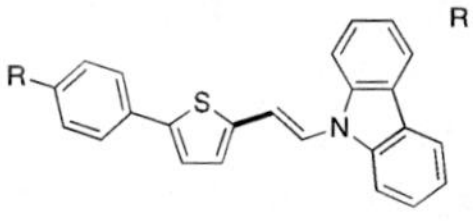

R =	No.	Yield
–H	**3**	74% (*E*:*Z* = > 20:1)
–NMe$_2$	**24**	78% (*E*:*Z* = > 20:1)
–OMe	**25**	77% (*E*:*Z* = > 20:1)
–F	**26**	77% (*E*:*Z* = > 20:1)
–CF$_3$	**27**	55% (*E*:*Z* = 14:1)

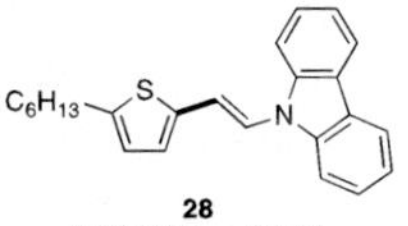

28
64% (*E*:*Z* = > 20:1)[a]

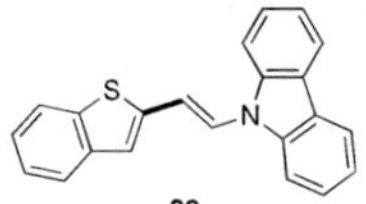

29
76% (*E*:*Z* = 15:1)

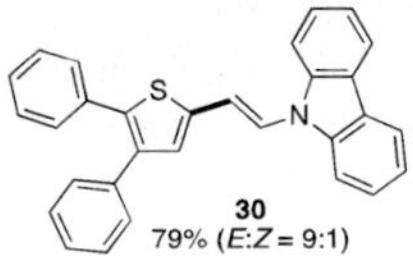

30
79% (*E*:*Z* = 9:1)

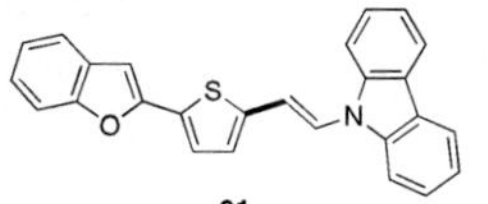

31
46% (*E*:*Z* = > 20:1)

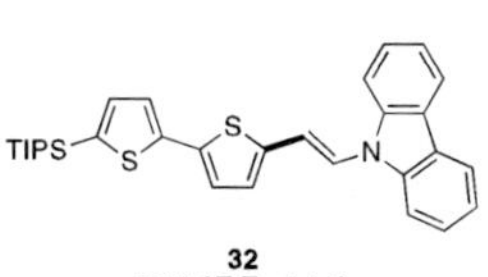

32
70% (*E*:*Z* = 14:1)

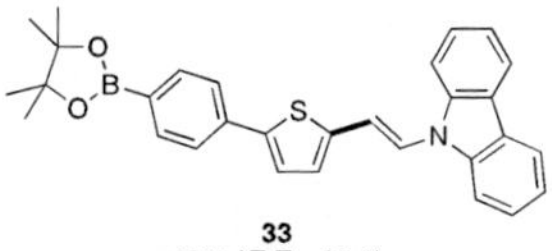

33
40% (*E*:*Z* = 15:1)

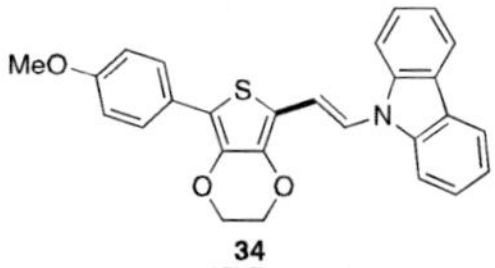

34
69% (*E*:*Z* = 15:1)

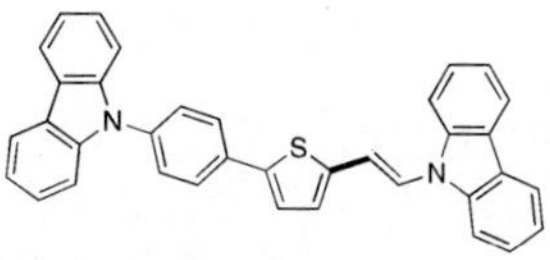

35
76% (*E*:*Z* = > 20:1)

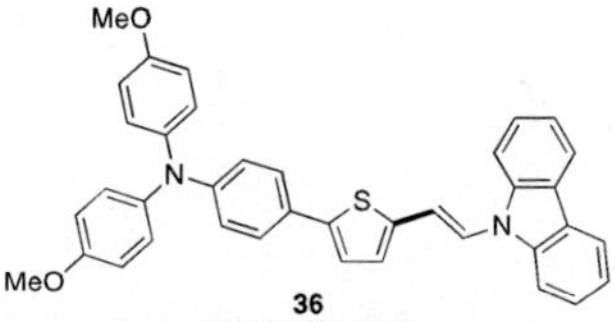

36
84% (*E*:*Z* = 15:1)
λ_{abs} = 400 nm, λ_{FL} = 507 nm, Φ_{FL} = 0.84

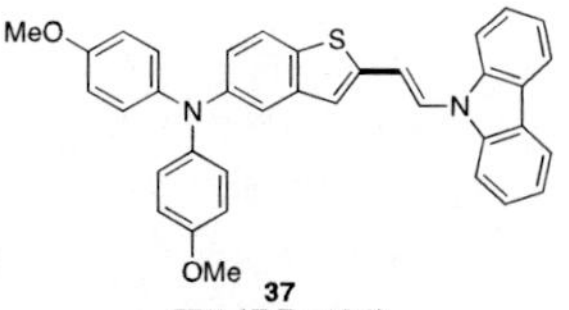

37
77% (*E*:*Z* = 12:1)

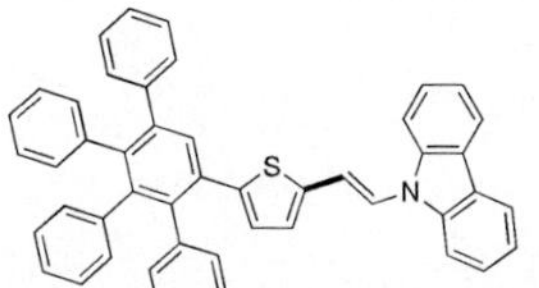

38
85% (*E*:*Z* = > 20:1)

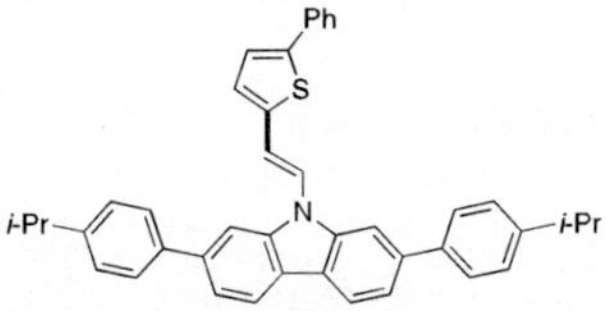

39
84% (*E*:*Z* = 7:1)[b]

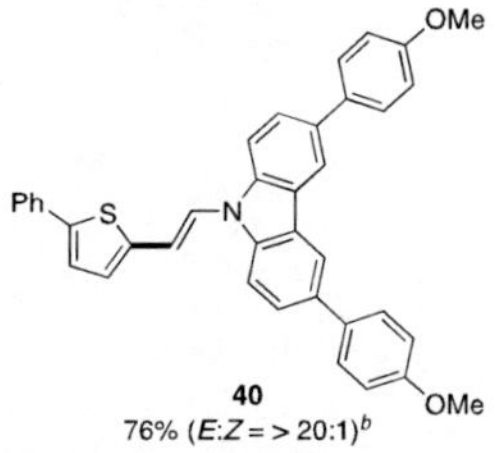

40
76% (*E*:*Z* = > 20:1)[b]

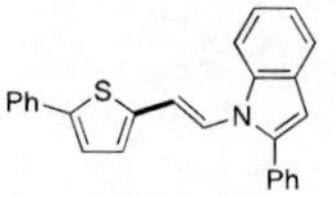

41
86% (*E*:*Z* = > 20:1)[b]

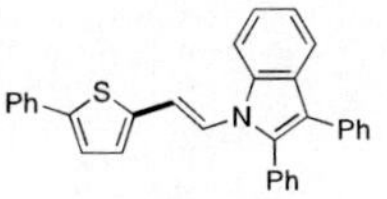

42
90% (*E*:*Z* = > 20:1)[b]

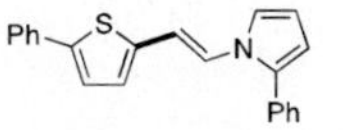

43
58% (*E*:*Z* = > 20:1)[b]

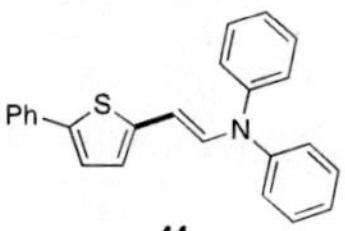

44
39% (*E*:*Z* = > 20:1)[b, c]

[a] With a thiophene (1.0 equiv), an enamine (2.0 equiv), Fe(acac)$_3$ (20 mol %), TP (22 mol %). [b] With a thiophene (1.5 equiv), an enamine (1.0 equiv), Fe(acac)$_3$ (20 mol %), TP (22 mol %). [c] At 70 °C.

(**46**) [18]. The IP value of **46** (–5.4 eV; film) was close to that of methylammonium lead iodide perovskite, indicating its use as a hole-transporting layer. Not only electron-rich thiophene monomers but also electron-deficient ones possessing ester or amide group were successfully incorporated into the polymer chain (**47**, **48**). 4,8-dialkoxybenzo[1,2-*b*:4,5-*b*′]dithiophene, a fused thiophene core, also afforded the product in high efficiency (**49**). A flat *N*,*N*'-divinylindolo[3,2-*b*]carbazole (ICZ) core reacted with slightly decreased efficiency (DP = 12), which may originate from the low solubility of the starting material and the product (**50**). The polymer chain may contain small amount of thiophene-thiophene homocoupling linkage judging from the formation of a homocoupling byproduct such as **4** in small molecule syntheses, although the exact ratio of homocoupling in polymer chain could not be determined either by ^{1}H or ^{13}C NMR due to the overlapped signals.

Table 5.6 Copolymerization of a bisthiophene with a bisenamine

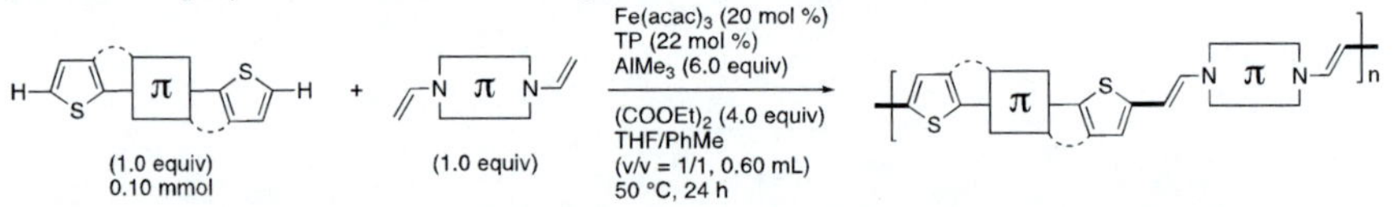

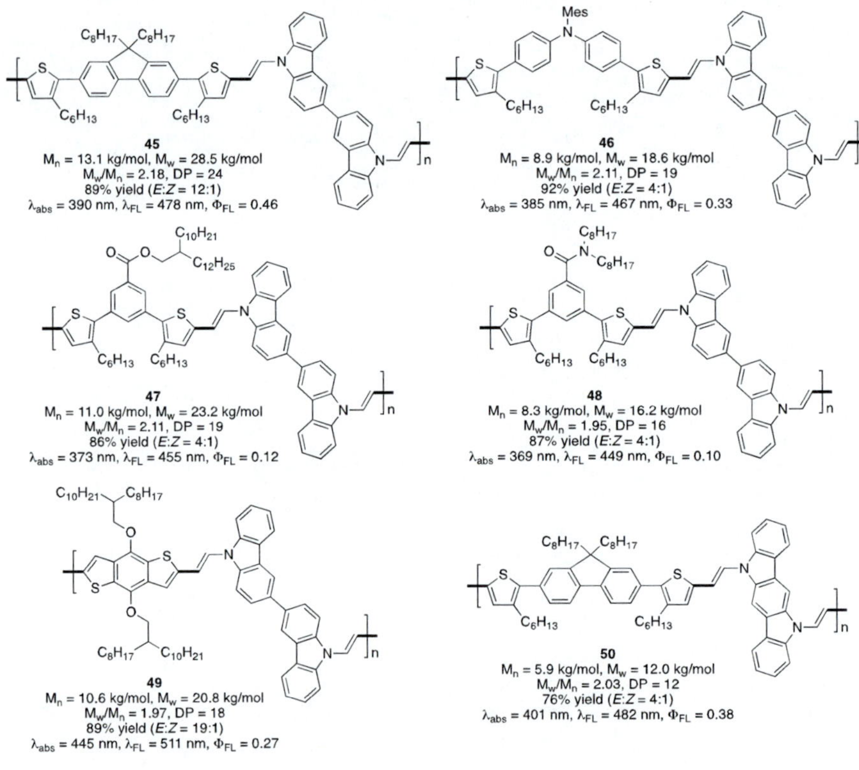

45
M_n = 13.1 kg/mol, M_w = 28.5 kg/mol
M_w/M_n = 2.18, DP = 24
89% yield (*E*:*Z* = 12:1)
λ_{abs} = 390 nm, λ_{FL} = 478 nm, Φ_{FL} = 0.46

46
M_n = 8.9 kg/mol, M_w = 18.6 kg/mol
M_w/M_n = 2.11, DP = 19
92% yield (*E*:*Z* = 4:1)
λ_{abs} = 385 nm, λ_{FL} = 467 nm, Φ_{FL} = 0.33

47
M_n = 11.0 kg/mol, M_w = 23.2 kg/mol
M_w/M_n = 2.11, DP = 19
86% yield (*E*:*Z* = 4:1)
λ_{abs} = 373 nm, λ_{FL} = 455 nm, Φ_{FL} = 0.12

48
M_n = 8.3 kg/mol, M_w = 16.2 kg/mol
M_w/M_n = 1.95, DP = 16
87% yield (*E*:*Z* = 4:1)
λ_{abs} = 369 nm, λ_{FL} = 449 nm, Φ_{FL} = 0.10

49
M_n = 10.6 kg/mol, M_w = 20.8 kg/mol
M_w/M_n = 1.97, DP = 18
89% yield (*E*:*Z* = 19:1)
λ_{abs} = 445 nm, λ_{FL} = 511 nm, Φ_{FL} = 0.27

50
M_n = 5.9 kg/mol, M_w = 12.0 kg/mol
M_w/M_n = 2.03, DP = 12
76% yield (*E*:*Z* = 4:1)
λ_{abs} = 401 nm, λ_{FL} = 482 nm, Φ_{FL} = 0.38

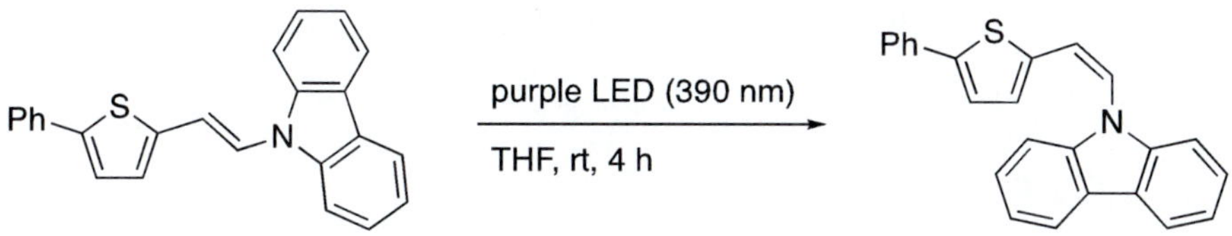

3-*E*
E:*Z* = > 20:1

3-*Z*
E:*Z* = 1: 14
99% yield

E/*Z* ratios were determined by ^{1}H NMR.

Scheme 5.10 EZ isomerization by light irradiation

5.8 *EZ* Isomerization

To further demonstrate the synthetic utility of this reaction, conversion of the *E* product to Z compound was attempted. By knowing that the energy of the S_1 state of compound **3** is 3.17 eV = 391 nm, the *E* product in THF was irradiated with purple LED (390 nm) at room temperature for 4 h. As expected, the Z product was obtained in excellent yield with high *Z* enrichment (Scheme 5.10).

5.9 Deuterium Labeling Experiments

To gain insight into the mechanism of C–H activation steps, I conducted deuterium labeling experiments. As shown in Scheme 5.11, one of the substrates was labeled with deuterium and the reaction was quenched at the middle point to determine the deuterium incorporation ratios. Interestingly, deuterium exchange was observed between two substrates. The reaction with deuterated 9-vinylcarbazole resulted in recovery of 34% deuterated 2-phenylthiophene (Scheme 5.11a) and the reaction with deuterated 2-phenylthiophene resulted in recovery of 18% deuterated 9-vinylcarbazole (Scheme 5.11b).

These phenomena can be explained by assuming reversible C–H activations between two substrates. Taking into account the low acidity of enamine [19], deuterium shift from enamine to thiophene (Scheme 5.11a) is well-explained by nucleophilic attack of electron-rich enamine to electron-deficient Fe(III) center, i.e., electrophilic metalation of enamine (Fig. 5.4a, **VI'** to **VII'**), followed by deprotonation by a thienyl group on iron (Fig. 5.4a, **VII'** to **IX**). Conversely, deuterium shift from thiophene to enamine (Scheme 5.11b) can be explained by σ-bond metathesis of the most acidic C–H bond of thiophene with enaminyl iron species (Fig. 5.4b, **IX to VI'**). Notably, deuterium exchange between two substrates would not take place if the reaction proceeds through a carbometallation/β-H elimination pathway because it would generate an iron hydride species that would not undergo hydride exchange with neither of the substrates (Fig. 5.4c). Therefore, deuterium exchange

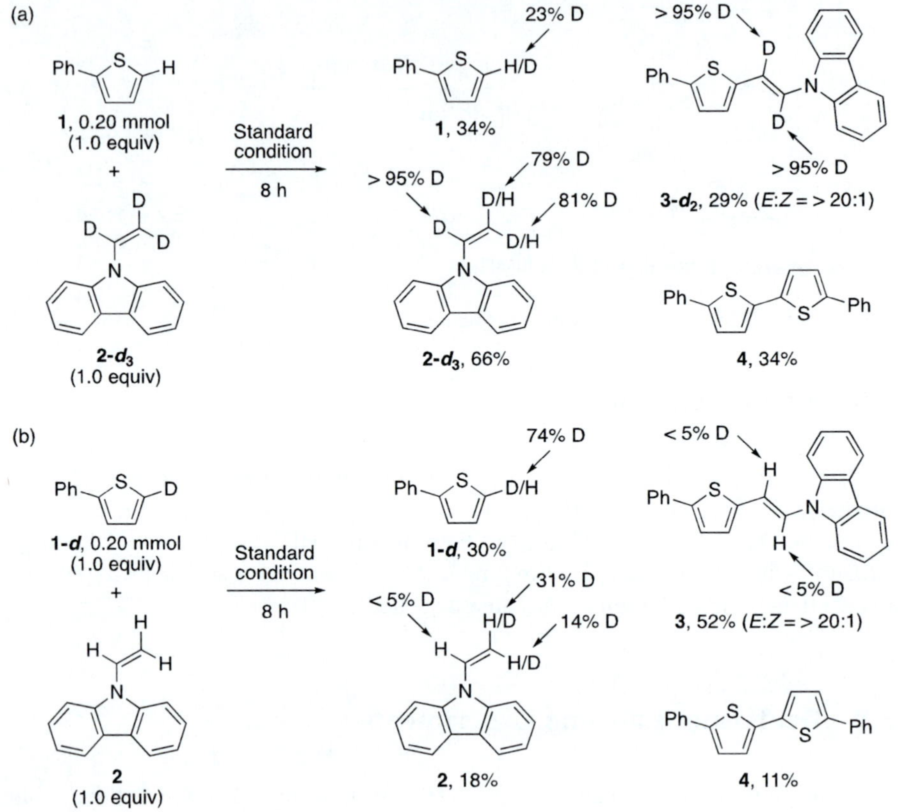

Scheme 5.11 Deuterium labeling experiments

between two substrates strongly supports the electrophilic metalation/deprotonation pathway (Fig. 5.2, **VI to VIII**) by intermediacy of **VII** (Fig. 5.2). Formation of the homocoupling byproduct **5** also supports the electrophilic metalation/deprotonation pathway rather than the carbometallation/β-H elimination pathway.

5.10 Kinetic Isotope Effect Experiments

Finally, kinetic isotope effect (KIE) experiments were conducted for two parallel reactions. To exclude the effect of deuterium exchange between two substrates, both of the substrates were labeled with deuterium (Scheme 5.12b). The yields of the cross-coupling product were traced up to 4 h, and the KIE value of 1.5 was obtained. This value is lower than the reported primary KIE values of 1.8 to 3.4 for iron-catalyzed C–H activation [20]. This indicates that neither of the C–H bond cleavage steps is the turnover-limiting step of the catalytic cycle [21]. The KIE value of 1.5 exceeds the

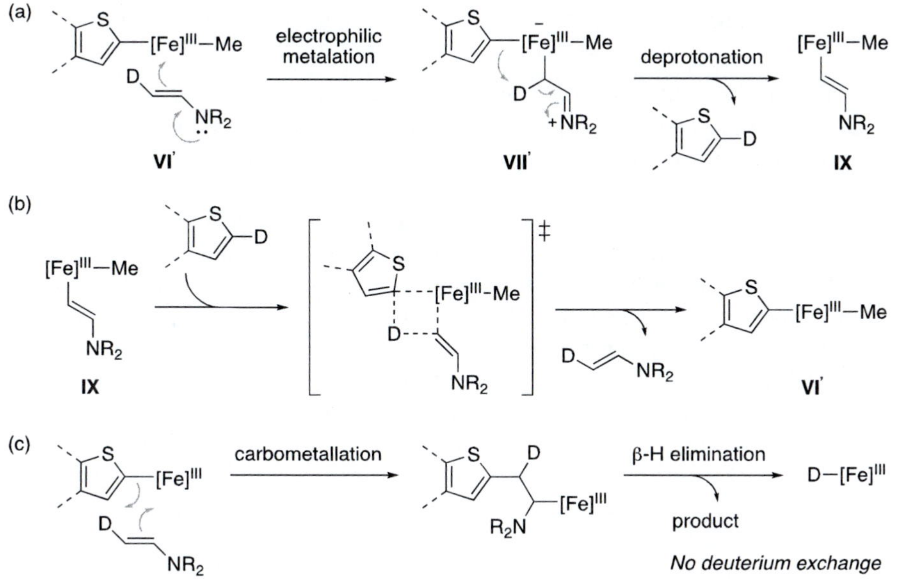

Fig. 5.4 Reversible C–H activation between two substrates

estimated maximum theoretical value of secondary KIE ($k_H/k_D = 1.4$) and therefore may be ascribed to the equilibrium isotope effect (EIE) of the precomplexation of 9-vinylcarbazole to the iron catalyst [22].

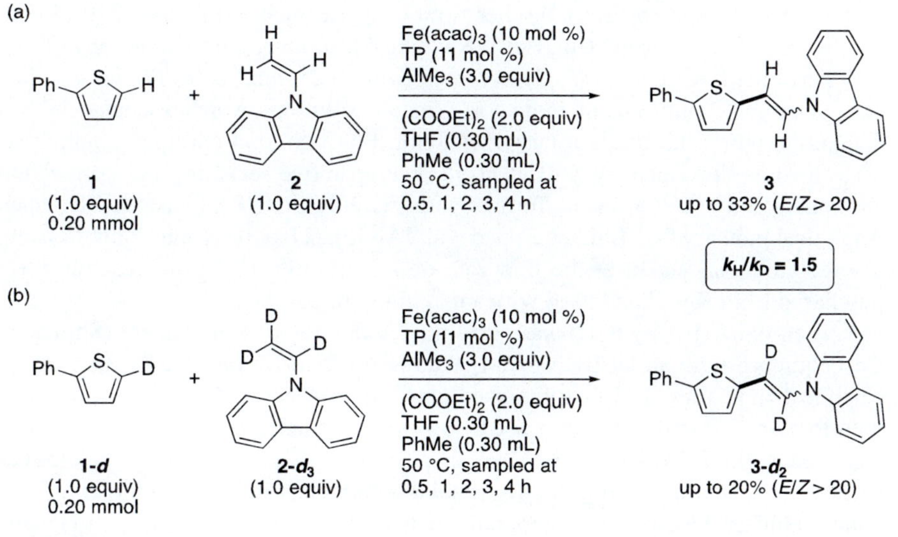

Scheme 5.12 Kinetic isotope effect experiments for two parallel reactions

5.11 Conclusion

In conclusion, iron-catalyzed oxidative C–H alkenylation, or Fujiwara–Moritani type reaction, was developed by applying the homocoupling reaction conditions to a mixture thiophenes and enamines. The reaction gave direct access to donor materials containing unique vinylthiophene-carbazole connected structures with controlled regioselectivity, branched/linear selectivity, and *E*/*Z* selectivity. Owing to the high efficiency, this reaction was applicable to the synthesis of copolymers containing a bisthiophene and a bisenamine monomer units. The obtained *E* product was isomerized to the corresponding *Z* compound by irradiation with light. Deuterium exchange between two substrates indicated the C–H activation takes place through an electrophilic metalation mechanism rather than a carbometallation mechanism. Kinetic isotope experiments indicated that neither of the C–H activation steps is the turnover-limiting step. This work demonstrates that Fe(III)/TP/$AlMe_3$/oxalate system is applicable to cross-coupling as well as homocoupling and paves the way to the development of iron-catalyzed C–H functionalization of thiophene compounds.

5.12 Experimental

Materials and methods

All air- or moisture-sensitive reactions were performed in a dry reaction vessel under argon atmosphere. Air- or moisture-sensitive liquids and solutions were transferred with syringe or Teflon cannula. The water content of solvents was confirmed to be less than 30 ppm by Karl Fischer titration performed with MKC-210 (Kyoto Electronics Manufacturing Co., Ltd.). Analytical thin-layer chromatography (TLC) was performed with a glass plate coated with 0.25 mm 230–400 mesh silica gel containing a fluorescent indicator. Organic solutions were evacuated with a diaphragm pump through a rotary evaporator. Flash column chromatography was performed as described by Still et al. [23]. Preparative recycling gel permeation chromatography (GPC) was performed with LC-92XX II NEXT instrument (Japan Analytical Industry Co., Ltd.) equipped with JAIGEL-2 h polystyrene columns using chloroform as an eluent at the flow rate of 7.5 mL/min. LEDs for reaction were purchased from Kessil and used with a maximum intensity.

Gas chromatography (GC) was performed with GC-2014 instrument (Shimadzu Co.) equipped with an ULBON HR-1 (0.25 mm I.D. $\times$ 25 mL, 0.25 μ m, Shinwa Chemical Industries, Ltd.) capillary column. Mass spectra (GC–MS) were taken with Parvum 2 instrument (Shimadzu Co.). High-resolution mass spectra (HRMS) were taken with LCMS-IT-TOF (Shimadzu Co.) using reserpine (MW 608.2734) as an internal standard. Melting points of solid compounds were measured on a Mel-Temp capillary melting-point apparatus and were uncorrected. Nuclear magnetic resonance (NMR) spectra were taken with ECZ-500 (JEOL, Ltd.) at room temperature unless otherwise noted and reported in parts per million (ppm). ^{1}H NMR spectra were internally referenced to tetramethylsilane (0.00 ppm), $CHCl_3$ (7.26 ppm), $CHDCl_2$ (5.32 ppm), C_2HDCl_4 (5.97 ppm), or $(CHD_2)(CD_3)SO$ (2.50 ppm). ^{13}C NMR spectra were internally referenced to tetramethylsilane (0.0 ppm), $CDCl_3$ (77.0 ppm), CD_2Cl_2 (53.8 ppm), $C_2D_2Cl_4$ (73.8 ppm), or $(CD_3)_2SO$ (39.5 ppm). ^{19}F NMR spectra were internally referenced to C_6F_6 (–164.9 ppm). ^{31}P NMR spectra were internally referenced to $(CH_3O)_3PO$ (2.1 ppm). ICP analysis was performed on Shimadzu ICPS-7510 equipment.

Unless otherwise noted, reagents were purchased from Tokyo Chemical Industry Co., Ltd., Sigma-Aldrich Co., LCC, FUJIFILM Wako Pure Chemical Co., and other commercial suppliers and were used as received. Anhydrous tetrahydrofuran and diethyl ether were purchased from KANTO Chemical Co., Inc. and purified prior to use by a solvent purification system (GlassContour) equipped with columns of activated alumina and supported copper catalyst [24]. $Fe(acac)_3$ (99.9% trace metal basis) was purchased from Sigma-Aldrich Co., LCC and used as received. Diethyl oxalate was purchased from Tokyo Chemical Industry Co., degassed by Freeze–Pump–Thaw cycling for three times, dried with molecular sieves 4A, and kept in a storage flask. Thiophene substrates were synthesized according to the literature [25].

Preparation of starting materials

2-Phenyl-1-vinyl-1*H*-indole:

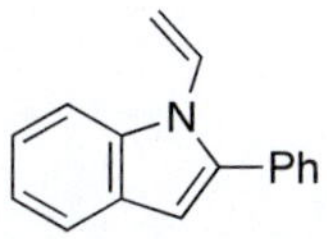

A mixture of 2-phenyl-1*H*-indole (0.97 g, 5.0 mmol), $t\text{-}Bu_3PH{\bullet}BF_4$ (58 mg, 0.20 mmol), $Pd(OAc)_2$ (23 mg, 0.10 mmol), Cs_2CO_3 (2.4 g, 7.5 mmol), and a THF solution of vinyl bromide (1 mol/L, 7.5 mL, 7.5 mmol) was stirred under argon atmosphere at 70 °C for 17 h. The reaction mixture was cooled to room temperature, diluted with toluene, and passed through a pad of Celite. The solvent was removed under reduced pressure, and the crude product was purified by silica gel chromatography (hexane: NEt_3 = 19:1) to afford the product as pale orange oil (0.86 g, 79%).

^1H NMR (500 MHz, $CDCl_3$): δ 7.72 (d, J = 8.9 Hz, 1H), 7.64 (d, J = 7.7 Hz, 1H), 7.56–7.54 (m, 2H), 7.47–7.38 (m, 3H), 7.29–7.27 (m, 1H), 7.20–7.17 (m, 1H), 6.96 (dd, J = 15.8, 9.0 Hz, 1H), 6.63 (s, 1H), 5.36 (d, J = 15.8 Hz, 1H), 5.08 (d, J = 9.0 Hz, 1H).

^{13}C NMR (125 MHz, $CDCl_3$): δ 140.4, 136.7, 132.5, 131.5, 129.5, 128.9, 128.4, 128.0, 122.6, 121.0, 120.7, 111.4, 104.7, 104.3.

HRMS (APCI +): *m/z* calcd for $C_{16}H_{13}N$ [M + H^+] 220.1121; found: 220.1115.

2-Phenylthiophene-5-*d*:

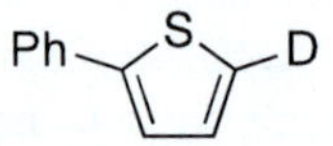

A hexane solution of BuLi (1.60 mol/L, 2.8 mL, 4.4 mmol) was added dropwise to a solution of 2-phenylthiophene (0.64 g, 4.0 mmol) in THF (8 mL) at –78 °C. After stirring for 1 h, D_2O (4 mL) was added dropwise and the reaction mixture was gradually warmed to room temperature and stirred for 30 min. The aqueous layer was extracted with diethyl ether (10 mL × 3), and the combined organic layers were dried over Na_2SO_4. The solvent was removed under reduced pressure, and the crude product was redissolved in hexane and passed through a pad of silica gel (hexane only) to afford the product as white solid (0.64 g, 100%). This compound is known [26].

^{1}H NMR (500 MHz, DMSO-d_6): δ 7.66–7.64 (m, 2H), 7.51 (d, $J = 3.7$ Hz, 1H), 7.42–7.39 (m, 2H), 7.31–7.28 (m, 1H), 7.13 (d, $J = 3.7$ Hz, 1H).

^{13}C NMR (125 MHz, DMSO-d_6): δ 143.2, 133.8, 129.1 (br), 128.4 (br), 127.6 (br), 125.8 (br), 125.4 (br), 123.7 (br) (Broad peaks were observed because of slow bond rotation.)

9-(Vinyl-*d*3)-9*H*-carbazole:

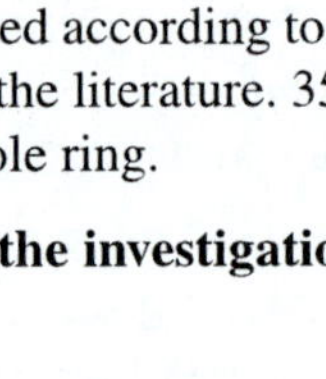

The title compound was synthesized according to the literature [27]. The compound data was in good agreement with the literature. 35% of deuterium was incorporated at 1 and 8 positions of the carbazole ring.

A representative procedure for the investigation of reaction parameters (Table 5.3)

In an oven-dried Schlenk tube was added 2-phenylthiophene (32 mg, 0.20 mmol), 9-vinyl-9*H*-carbazole (58 mg, 0.30 mmol), TP (14 mg, 0.022 mmol), and a THF solution of $Fe(acac)_3$ (0.067 mol/L, 0.30 mL, 0.020 mmol). Then, a toluene solution of $AlMe_3$ (2.0 mol/L, 0.30 mL, 0.60 mmol) was added by rinsing the wall of the Schlenk tube at rt. Diethyl oxalate (54 μ L, 0.40 mmol) was added, and the reaction mixture was stirred at 70 °C for 15 h. The reaction mixture was cooled to rt, diluted with ethyl acetate (1 mL), and quenched carefully with methanol (0.1 mL) and a saturated aqueous solution of potassium sodium tartrate (1 mL). The reaction mixture was diluted with toluene until all the products are dissolved and was stirred vigorously until clear phase separation was observed. Tridecane (30 μ L) was added as an internal standard, and a portion of the organic layer was passed through a pad of Florisil and analyzed by GC.

A general procedure for iron-catalyzed oxidative C–H alkenylation of thiophenes with enamines (Table 5.5)

In an oven-dried Schlenk tube was added a thiophene (0.40 mmol), an enamine (0.60 mmol), TP (28 mg, 0.044 mmol), and a THF solution of $Fe(acac)_3$ (0.067 mol/L, 0.60 mL, 0.040 mmol). Then, a toluene solution of $AlMe_3$ (2.0 mol/L, 0.60 mL, 1.2 mmol) was added by rinsing the wall of the Schlenk tube. Diethyl oxalate (108 μ L, 0.80 mmol) was added, and the reaction mixture was stirred at 50 °C for 15 h. The reaction mixture was cooled to rt, diluted with toluene (2 mL), and quenched carefully with methanol (0.1 mL) and a saturated aqueous solution of potassium sodium tartrate (2 mL). The reaction mixture was diluted with toluene until the product was dissolved and was stirred vigorously until clear phase separation was observed. The aqueous layer was extracted either with toluene (5 mL × 3) or dichloromethane (if there are any insoluble byproducts), and the combined organic layers were passed through a pad of Celite and Florisil. The solvent was removed under reduced pressure, and the crude product was purified by silica gel chromatography using Isolera (hexane/dichloromethane gradient). *E*/*Z* ratios were determined by integration of the ^{1}H NMR spectra.

(*E*)-9-(2-(5-phenylthiophen-2-yl)vinyl)-9*H*-carbazole (3):

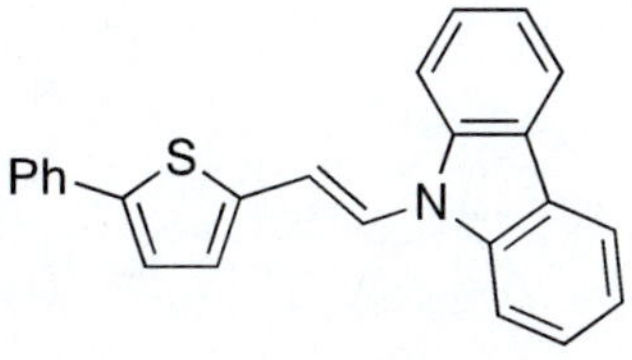

The title compound was obtained as pale yellow solid in 74% yield with $E/Z > 20$. The reaction was performed on a 0.40 mmol scale using 1.5 equiv of the enamine.

Melting point: 156–157 °C (dichloromethane).

^{1}H NMR (500 MHz, $CDCl_3$): δ 8.11–8.09 (m, 2H), 7.74–7.72 (m, 2H), 7.67–7.64 (m, 3H), 7.53–7.50 (m, 2H), 7.43–7.40 (m, 2H), 7.34–7.28 (m, 3H), 7.27 (d, J = 3.7 Hz, 1H), 7.18 (d, J = 14.3 Hz, 1H), 7.05 (d, J = 3.7 Hz, 1H).

^{13}C NMR (125 MHz, $CDCl_3$): δ 142.1, 139.9, 139.4, 134.2, 129.0, 127.5, 126.4, 126.3, 125.6, 124.1, 123.6, 122.6, 120.9, 120.4, 113.5, 110.6.

HRMS (APCI +): m/z calcd for $C_{24}H_{17}NS$ [$M + H^+$] 352.1154; found: 352.1147.

(*E*)-4-(5-(2-(9*H*-carbazol-9-yl)vinyl)thiophen-2-yl)-*N*,*N*-dimethylaniline (24):

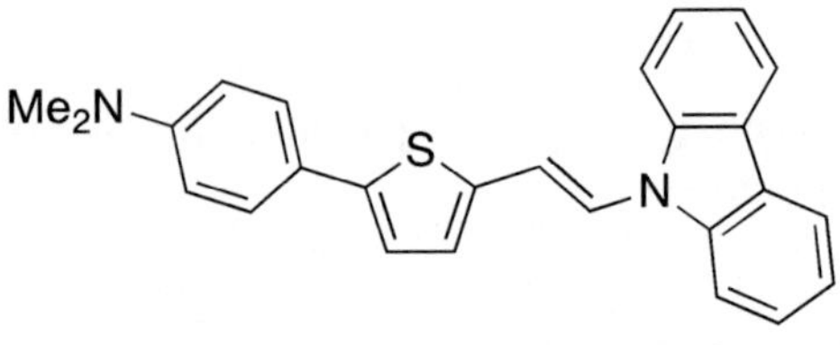

The title compound was obtained as yellow solid in 78% yield with *E*/*Z* > 20. The reaction was performed on a 0.40 mmol scale using 1.5 equiv of the enamine.

Melting point: 175–176 °C (dichloromethane).

^{1}H NMR (500 MHz, C_6D_6): δ 7.54 (d, J = 7.8 Hz, 2H), 7.26 (m, 2H), 7.02 (d, J = 7.7 Hz, 2H), 6.96 (d, J = 14.3 Hz, 1H), 6.92–6.88 (m, 2H), 6.82–6.79 (m, 2H), 6.67 (d, J = 3.6 Hz, 1H), 6.50 (d, J = 14.3 Hz, 1H), 6.36 (d, J = 3.6 Hz, 1H), 6.16–6.16 (m, 2H), 2.07 (s, 6H).

^{13}C NMR (125 MHz, C_6D_6): δ 150.3, 144.0, 139.9, 137.9, 127.0 (two signals overlapped), 126.5, 124.5, 123.3, 121.9, 121.5, 121.0, 120.5, 115.1, 112.9, 111.0, 39.9.

HRMS (APCI +): *m/z* calcd for $C_{26}H_{22}N_2S$ [M + H$^+$] 395.1576; found: 395.1571.

(*E*)-9-(2-(5-(4-methoxyphenyl)thiophen-2-yl)vinyl)-9*H*-carbazole (25):

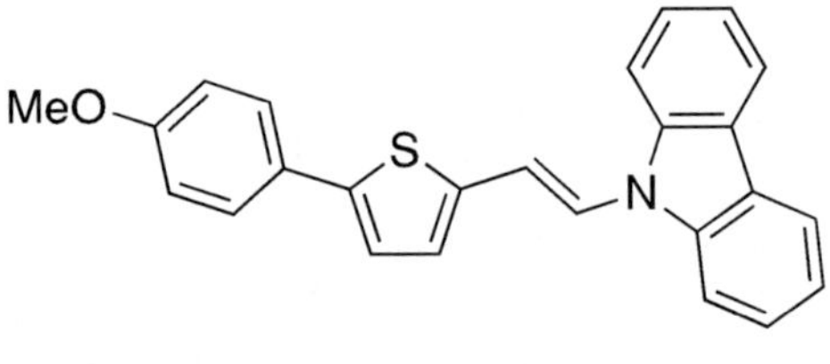

The title compound was obtained as pale yellow solid in 77% yield with *E*/*Z* > 20. The reaction was performed on a 0.40 mmol scale using 1.5 equiv of the enamine.

Melting point: 153–155 °C (dichloromethane).

^{1}H NMR (500 MHz, $CDCl_3$): δ 8.10 (d, J = 7.8 Hz, 2H), 7.73 (d, J = 8.3 Hz, 2H), 7.62 (d, J = 14.3 Hz, 1H), 7.58–7.56 (m, 2H), 7.53–7.49 (m, 2H), 7.34–7.31 (m, 2H), 7.18–7.15 (m, 2H), 7.02 (d, J = 3.7 Hz, 1H), 6.95–6.94 (m, 2H), 3.85 (s, 3H).

^{13}C NMR (125 MHz, $CDCl_3$): δ 159.2, 142.1, 139.4, 138.8, 127.0, 126.9, 126.3, 126.3, 124.1, 122.5, 122.2, 120.8, 120.3, 114.3, 113.9, 110.6, 55.4.

HRMS (APCI +): *m/z* calcd for $C_{25}H_{19}NOS$ [M + H$^+$] 382.1260; found: 382.1269.

(*E*)-9-(2-(5-(4-fluorophenyl)thiophen-2-yl)vinyl)-9*H*-carbazole (26):

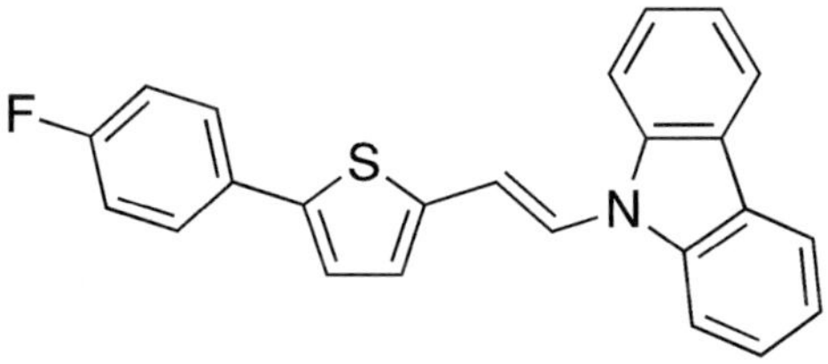

The title compound was obtained as pale yellow solid in 77% yield with $E/Z > 20$. The reaction was performed on a 0.40 mmol scale using 1.5 equiv of the enamine.

Melting point: 145–146 °C (dichloromethane).

^{1}H NMR (500 MHz, $CDCl_3$): δ 8.10 (d, J = 7.5 Hz, 2H), 7.73 (d, J = 7.7 Hz, 2H), 7.65 (d, J = 14.3 Hz, 1H), 7.61–7.58 (m, 2H), 7.53–7.50 (m, 2H), 7.35–7.31 (m, 2H), 7.19 (d, J = 3.6 Hz, 1H), 7.17 (d, J = 14.3 Hz, 1H), 7.12–7.09 (m, 2H), 7.03 (d, J = 3.6 Hz, 1H).

^{13}C NMR (125 MHz, $CDCl_3$): δ 162.3 (d, J = 247 Hz), 140.9, 139.9, 139.4, 130.5 (d, J = 3.6 Hz), 127.2 (d, J = 7.8 Hz), 126.4, 126.2, 124.2, 123.6, 122.7, 121.0, 120.4, 115.9 (d, J = 21.7 Hz), 113.3, 110.6.

^{19}F NMR (471 MHz, $CDCl_3$): δ–117.6.

HRMS (APCI +): *m/z* calcd for $C_{24}H_{16}FNS$ [M + H^+] 370.1060; found: 370.1053.

(*E*)-9-(2-(5-(4-(trifluoromethyl)phenyl)thiophen-2-yl)vinyl)-9*H*-carbazole (27):

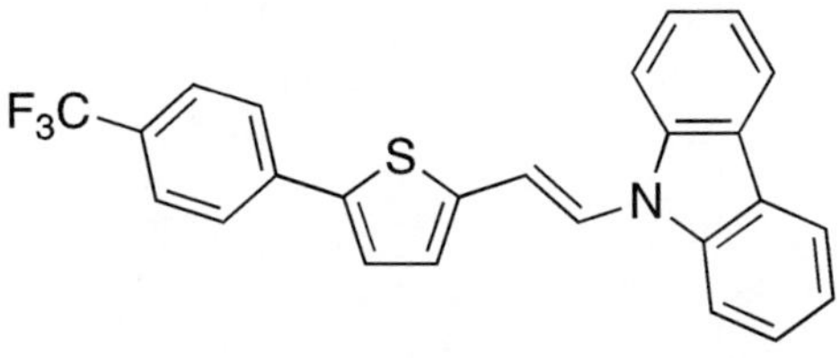

The title compound was obtained as yellow solid in 55% yield with $E/Z = 14$. The reaction was performed on a 0.40 mmol scale using 1.5 equiv of the enamine.

Melting point: 166–169 °C (dichloromethane).

^{1}H NMR (500 MHz, $CDCl_3$): δ 8.11–8.09 (m, 2H), 7.74–7.72 (m, 4H), 7.69–7.64 (m, 3H), 7.54–7.50 (m, 2H), 7.36–7.32 (m, 3H), 7.18 (d, J = 14.3 Hz, 1H), 7.08 (d, J = 3.8 Hz, 1H).

^{13}C NMR (125 MHz, $CDCl_3$): δ 141.5, 139.9, 139.3, 137.5, 126.4, 126.2, 126.0 (q, J = 3.6 Hz), 125.5, 125.1, 124.3, 124.1 (q, J = 272 Hz), 123.4, 121.1, 120.4, 112.5, 110.9, 110.6.

^{19}F NMR (471 MHz, $CDCl_3$): δ–65.7.

HRMS (APCI +): *m/z* calcd for $C_{25}H_{16}F_3NS$ [M + H^+] 420.1028; found: 420.1027.

(*E*)-9-(2-(5-hexylthiophen-2-yl)vinyl)-9*H*-carbazole (28):

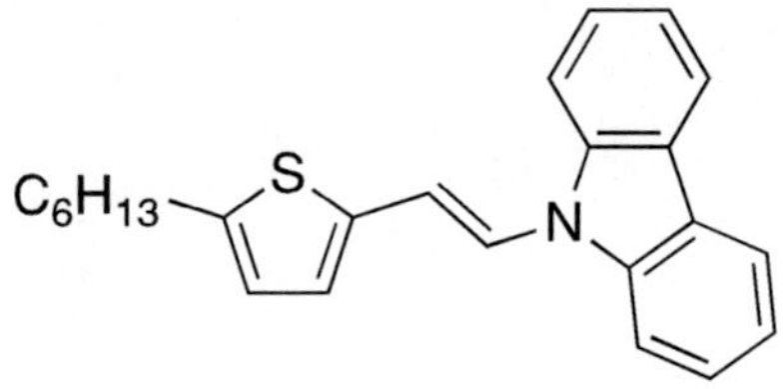

The title compound was obtained as colorless oil in 64% yield with *E*:*Z* = > 20:1. The reaction was performed on a 0.40 mmol scale using 1.0 equiv of the thiophene, 2.0 equiv of the enamine, 20 mol % of $Fe(acac)_3$, and 22 mol % of TP.

^{1}H NMR (500 MHz, $CDCl_3$): δ 8.09 (d, J = 7.4 Hz, 2H), 7.69 (d, J = 8.3 Hz, 2H), 7.51–7.47 (m, 3H), 7.32–7.29 (m, 2H), 7.12 (d, J = 14.1 Hz, 1H), 6.88 (d, J = 3.4 Hz, 1H), 6.71 (d, J = 3.4 Hz, 1H), 2.83 (t, J = 7.5 Hz, 2H), 1.74–1.68 (m, 2H), 1.42–1.32 (m, 6H), 0.91 (t, J = 7.2 Hz, 3H).

^{13}C NMR (125 MHz, $CDCl_3$): δ 144.7, 139.5, 137.7, 126.3, 125.2, 124.6, 123.9, 121.5, 120.7, 120.3, 114.8, 110.5, 31.6, 31.6, 30.3, 28.8, 22.6, 14.1.

HRMS (APCI +): *m/z* calcd for $C_{24}H_{25}NS$ [M + H^+] 360.1780; found: 360.1771.

(*E*)-9-(2-(benzo[*b*]thiophen-2-yl)vinyl)-9*H*-carbazole (29):

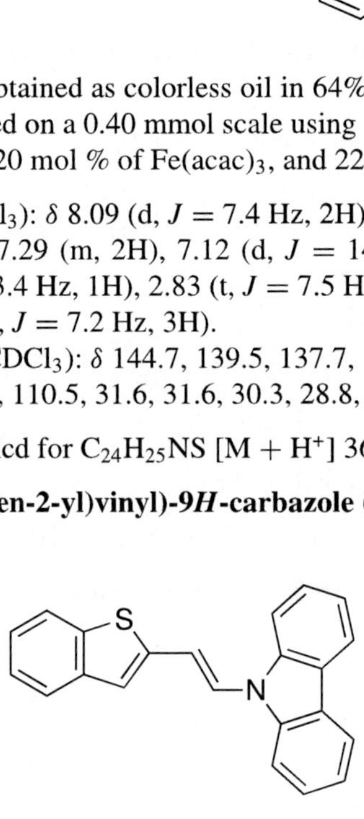

The title compound was obtained as white solid in 76% yield with *E*:*Z* = 15:1. The reaction was performed on a 0.40 mmol scale using 1.0 equiv of the thiophene and 1.5 equiv of the enamine.

Melting point: 127–128 °C (dichloromethane).

^{1}H NMR (500 MHz, $CDCl_3$): δ 8.11–8.09 (m, 2H), 7.82–7.80 (m, 1H), 7.76–7.70 (m, 4H), 7.54–7.51 (m, 2H), 7.34–7.29 (m, 5H). 7.26 (s, 1H).

^{13}C NMR (125 MHz, $CDCl_3$): δ 140.7, 140.3, 139.3, 138.3, 126.5, 124.7, 124.6, 124.5, 124.3, 123.1, 122.2, 121.9, 121.2, 120.4, 112.8, 110.7.

HRMS (APCI +): *m/z* calcd for $C_{22}H_{15}NS$ [M + H^+] 326.0998; found: 326.0992.

(*E*)-9-(2-(4,5-diphenylthiophen-2-yl)vinyl)-9*H*-carbazole (30):

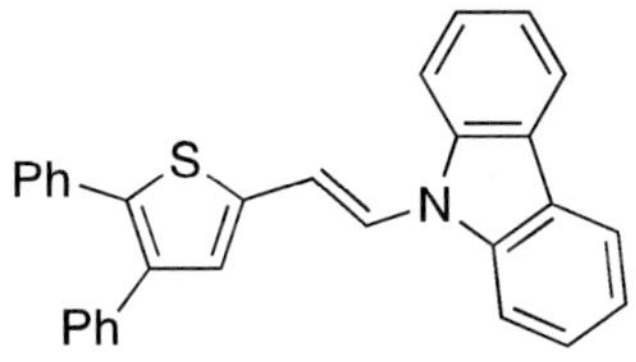

The title compound was obtained as yellow oil in 79% yield with *E*:*Z* = 9:1. The reaction was performed on a 0.20 mmol scale using 1.0 equiv of the thiophene and 1.5 equiv of the enamine.

^{1}H NMR (500 MHz, CD_2Cl_2): δ 8.04–8.02 (m, 2H), 7.70–7.68 (m, 2H), 7.64 (d, J = 14.3 Hz, 1H), 7.46–7.43 (m, 2H), 7.27–7.18 (m, 12H), 7.13 (d, J = 14.3 Hz, 1H), 7.10 (s, 1H).

^{13}C NMR (125 MHz, CD_2Cl_2): δ 139.7, 139.3, 139.0, 136.7, 136.6, 134.5, 129.4 (br), 129.3 (br), 128.9 (br), 128.8 (br), 128.8 (br), 127.8 (br), 127.4 (br), 126.7, 124.4, 123.2, 121.3, 120.6, 113.0 (br), 111.0. (Broad signals were observed because of slow rotation of bonds.)

HRMS (APCI +): *m/z* calcd for $C_{30}H_{21}NS$ [M + H$^+$] 428.1467; found: 428.1459.

(*E*)-9-(2-(5-(benzofuran-2-yl)thiophen-2-yl)vinyl)-9*H*-carbazole (31):

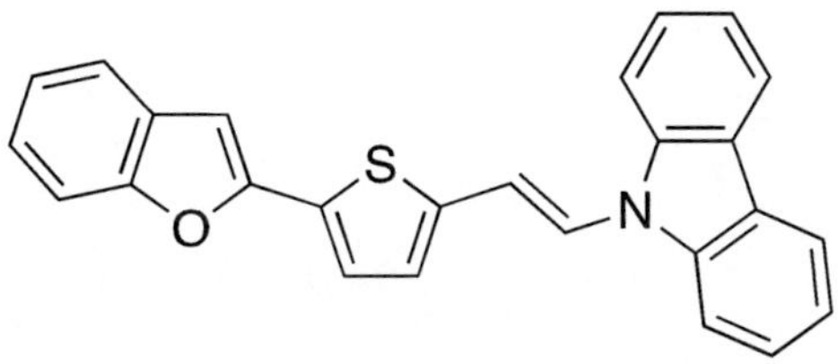

The title compound was obtained as yellow solid in 46% yield with *E*/*Z* > 20. The reaction was performed on a 0.20 mmol scale using 1.5 equiv of the enamine.

Melting point: 183–185 °C (dichloromethane).

^{1}H NMR (500 MHz, CD_2Cl_2): δ 8.04–8.03 (m, 2H), 7.70–7.65 (m, 3H), 7.51–7.50 (m, 1H), 7.47–7.42 (m, 3H), 7.37 (d, J = 3.7 Hz, 1H), 7.28–7.13 (m, 5H), 7.03 (d, J = 3.7 Hz, 1H), 6.85 (d, J = 0.9 Hz, 1H).

^{13}C NMR (125 MHz, CD_2Cl_2): δ 154.9, 151.4, 141.6, 139.6, 130.7, 129.5, 126.8, 126.3, 125.6, 124.8, 124.5, 123.7, 123.6, 121.4, 121.1, 120.7, 112.6, 111.2, 111.0, 101.5.

HRMS (APCI +): *m/z* calcd for $C_{26}H_{17}NOS$ [M + H$^+$] 392.1104; found: 392.1108.

(*E*)-9-(2-(5'-(triisopropylsilyl)-[2,2'-bithiophen]-5-yl)vinyl)-9*H*-carbazole (32):

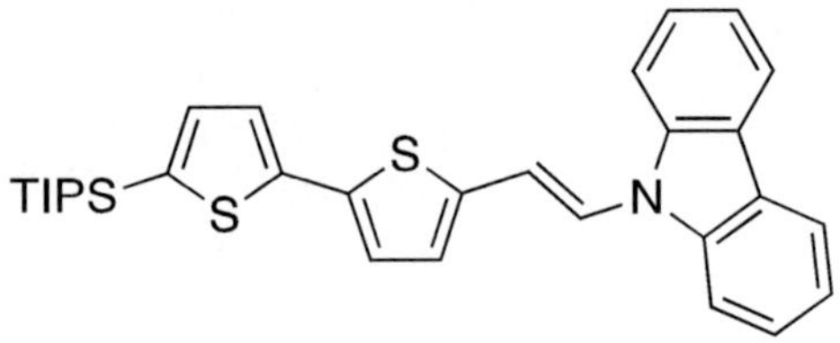

The title compound was obtained as yellow oil in 70% yield with E:Z = 14:1. The reaction was performed on a 0.20 mmol scale using 1.0 equiv of the thiophene and 1.5 equiv of the enamine.

^{1}H NMR (500 MHz, $CDCl_3$): δ 8.10–8.08 (m, 2H), 7.72 (d, J = 8.3 Hz, 2H), 7.61 (d, J = 14.3 Hz, 1H), 7.53–7.49 (m, 2H), 7.34–7.31 (m, 2H), 7.29 (d, J = 3.5 Hz, 1H), 7.18 (d, J = 3.5 Hz, 1H), 7.16–7.13 (m, 2H), 6.97 (d, J = 3.8 Hz, 1H), 1.36 (sep, J = 7.5 Hz, 3H), 1.14 (d, J = 7.5 Hz, 18H).

^{13}C NMR (125 MHz, $CDCl_3$): δ 142.2, 139.3, 139.3, 136.5, 135.3, 133.9, 126.4, 126.1, 124.6, 124.1, 124.1, 122.5, 120.9, 120.4, 113.2, 110.6, 18.6, 11.8.

HRMS (APCI+): m/z calcd for $C_{31}H_{35}NS_2Si$ [M + H$^+$] 514.2053; found: 514.2036.

(*E*)-9-(2-(5-(4-(4,4,5,5-tetramethyl-1,3,2-dioxaborolan-2-yl)phenyl)thiophen-2-yl)vinyl)-9*H*-carbazole (33):

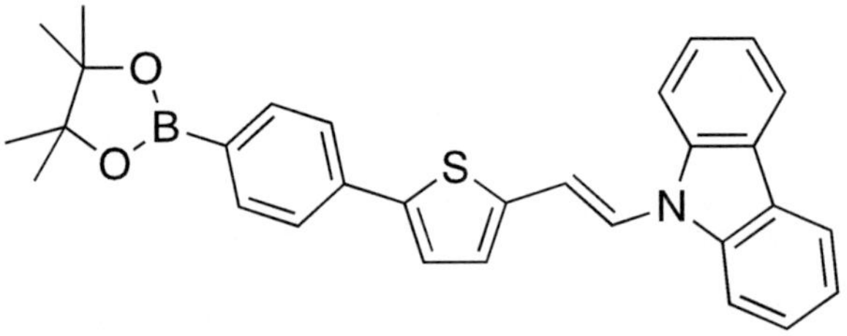

The title compound was obtained as yellow solid in 40% yield with E:Z = 15:1. The reaction was performed on a 0.20 mmol scale using 1.0 equiv of the thiophene and 1.5 equiv of the enamine.

Melting point: 186–188 °C (dichloromethane).

^{1}H NMR (500 MHz, $CDCl_3$): δ 8.09 (d, J = 7.7 Hz, 2H), 7.85–7.83 (m, 2H), 7.73 (d, J = 8.3 Hz, 2H), 7.68–7.63 (m, 3H), 7.53–7.50 (m, 2H), 7.34–7.31 (m, 3H), 7.17 (d, J = 14.3 Hz, 1H), 7.05 (d, J = 3.7 Hz, 1H), 1.37 (s, 12H).

^{13}C NMR (125 MHz, $CDCl_3$): δ 141.8, 140.5, 139.3, 136.6, 135.4, 126.4, 126.3, 124.7, 124.2, 124.2, 122.8, 121.0, 120.4, 113.2, 110.6, 83.9, 24.9. (*C*–B carbon signal is missing.)

HRMS (APCI +): *m/z* calcd for $C_{30}H_{28}BNO_2S$ [M + H$^+$] 478.2012; found: 478.2021.

(*E*)-9-(2-(7-(4-methoxyphenyl)-2,3-dihydrothieno[3,4-*b*][1,4]dioxin-5-yl)vinyl)-9*H*-carbazole (34):

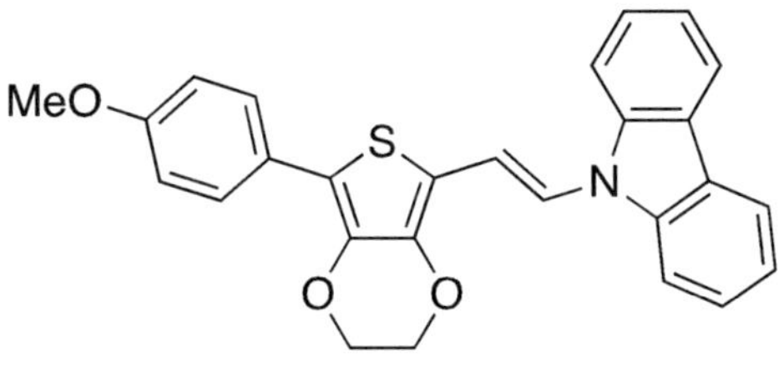

The title compound was obtained as yellow oil in 69% yield with *E*:*Z* = 15:1. The reaction was performed on a 0.20 mmol scale using 1.0 equiv of the thiophene and 1.5 equiv of the enamine.

^{1}H NMR (500 MHz, CD_2Cl_2): δ 8.02 (d, *J* = 7.8 Hz, 2H), 7.66 (d, *J* = 8.3 Hz, 2H), 7.60–7.56 (m, 3H), 7.44–7.41 (m, 2H), 7.24–7.21 (m, 2H), 6.99 (d, *J* = 14.4 Hz, 1H), 6.85–6.83 (m, 2H), 4.30–4.27 (m, 4H), 3.74 (s, 3H).

^{13}C NMR (125 MHz, CD_2Cl_2): δ 158.8, 139.8, 139.6, 138.0, 127.5, 126.6, 126.0, 124.2, 121.3, 121.0, 120.5, 114.4, 111.3, 110.9, 110.7, 65.3, 65.2, 55.6. (one signal overlapped).

HRMS (APCI +): *m/z* calcd for $C_{27}H_{21}NO_3S$ [M + H$^+$] 440.1315; found: 440.1322.

(*E*)-9-(2-(5-(4-(9*H*-carbazol-9-yl)phenyl)thiophen-2-yl)vinyl)-9*H*-carbazole (35):

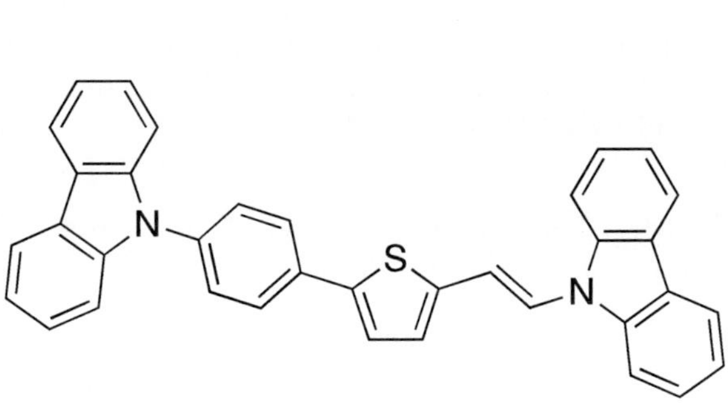

The title compound was obtained as yellow solid in 76% yield with E:$Z \geq 20$:1. The reaction was performed on a 0.20 mmol scale using 1.0 equiv of the thiophene and 1.5 equiv of the enamine.

Melting point: 230–232 °C (dichloromethane).

^{1}H NMR (500 MHz, $CDCl_3$): δ 8.17 (d, J = 7.7 Hz, 2H), 8.11 (d, J = 7.8 Hz, 2H), 7.87–7.86 (m, 2H), 7.77 (d, J = 8.0 Hz, 2H), 7.72 (d, J = 14.3 Hz, 1H), 7.63–7.61 (m, 2H), 7.55–7.42 (m, 6H), 7.37 (d, J = 3.7 Hz, 1H), 7.36–7.30 (m, 4H), 7.22 (d, J = 14.3 Hz, 1H), 7.11 (d, J = 3.7 Hz, 1H).

^{13}C NMR (125 MHz, CD_2Cl_2): δ 141.2, 141.0, 140.9, 139.7, 137.1, 133.6, 127.7, 127.1, 126.8, 126.8, 126.4, 124.6, 124.5, 123.7, 123.3, 121.4, 120.7, 120.6, 120.4, 113.2, 111.0, 110.1.

HRMS (APCI +): m/z calcd for $C_{36}H_{24}N_2S$ [M + H^+] 517.1733; found: 517.1719.

(*E*)-4-(5-(2-(9*H*-carbazol-9-yl)vinyl)thiophen-2-yl)-*N*,*N*-bis(4-methoxyphenyl)aniline (36):

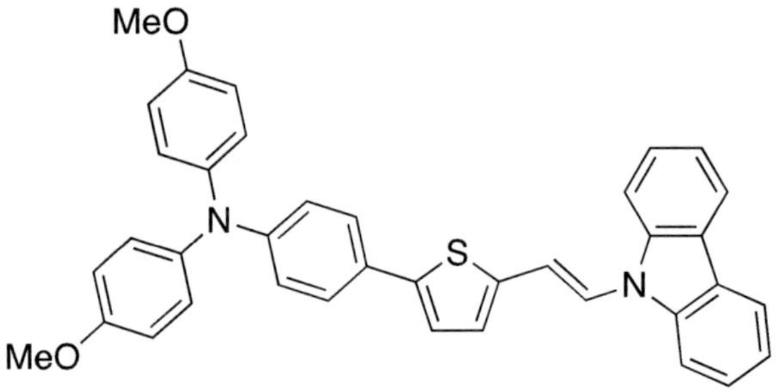

The title compound was obtained as yellow oil in 84% yield with E:Z = 15:1. The reaction was performed on a 0.20 mmol scale using 1.0 equiv of the thiophene and 1.5 equiv of the enamine.

^{1}H NMR (500 MHz, CD_2Cl_2): δ 8.10 (d, J = 7.8 Hz, 2H), 7.73 (d, J = 8.3 Hz, 2H), 7.62 (d, J = 14.3 Hz, 1H), 7.52–7.49 (m, 2H), 7.44–7.42 (m, 2H), 7.33–7.30 (m, 2H), 7.18 (d, J = 14.3 Hz, 1H), 7.13 (d, J = 3.7 Hz, 1H), 7.08–7.06 (m, 4H), 7.03 (d, J = 3.7 Hz, 1H), 6.90–6.88 (m, 2H), 6.86–6.84 (m, 4H), 3.78 (s, 6H).

^{13}C NMR (125 MHz, CD_2Cl_2): δ 156.6, 148.8, 142.7, 140.8, 139.7, 138.7, 127.2, 126.9, 126.7, 126.4, 126.3, 124.3, 122.4, 122.4, 121.2, 120.6, 120.4, 115.0, 114.1, 110.9, 55.8.

HRMS (APCI+): m/z calcd for $C_{38}H_{30}N_2O_2S$ [M + H^+] 579.2101; found: 579.2114.

(*E*)-2-(2-(9*H*-carbazol-9-yl)vinyl)-*N*,*N*-bis(4-methoxyphenyl)benzo[*b*]thiophen-5-amine (37):

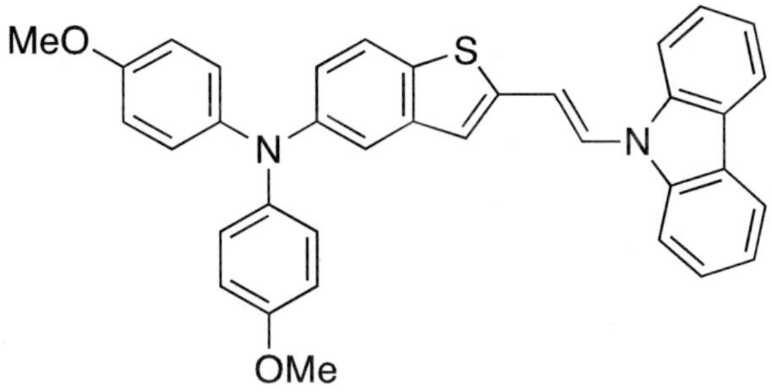

The title compound was obtained as yellow oil in 77% yield with *E*:*Z* = 12:1. The reaction was performed on a 0.20 mmol scale using 1.0 equiv of the thiophene and 1.5 equiv of the enamine.

^{1}H NMR (500 MHz, CD_2Cl_2): δ 8.02 (d, J = 7.5 Hz, 2H), 7.69 (d, J = 8.3 Hz, 2H), 7.65 (d, J = 14.3 Hz, 1H), 7.51 (d, J = 8.7 Hz, 1H), 7.46–7.43 (m, 2H), 7.27–7.24 (m, 2H), 7.18–7.16 (m, 2H), 6.99 (s, 1H), 6.97–6.96 (m, 4H), 6.91 (dd, J = 8.7, 2.3 Hz, 1H), 6.76–6.75 (m, 4H), 3.70 (s, 6H).

^{13}C NMR (125 MHz, CD_2Cl_2): δ 156.0, 146.9, 141.9, 141.8, 141.7, 139.6, 131.3, 126.8, 126.3, 124.9, 124.6, 122.7, 122.0, 121.5, 120.8, 120.6, 115.7, 114.9, 113.0, 111.1, 55.8.

HRMS (APCI +): *m/z* calcd for $C_{36}H_{28}N_2O_2S$ [M + H$^+$] 553.1944; found: 553.1947.

(*E*)-9-(2-(5-(3',6'-diphenyl-[1,1':2',1''-terphenyl]-4'-yl)thiophen-2-yl)vinyl)-9*H*-carbazole (38):

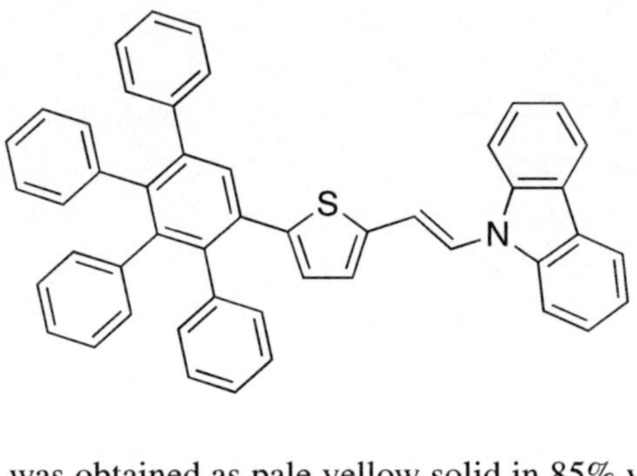

The title compound was obtained as pale yellow solid in 85% yield with *E*/*Z* > 20. The reaction was performed on a 0.20 mmol scale using 1.5 equiv of the enamine.

Melting point: 262–264 °C (dichloromethane).

^{1}H NMR (500 MHz, $CDCl_3$): δ 8.07–8.06 (m, 2H), 7.79 (s, 1H), 7.66–7.64 (m, 2H), 7.50–7.46 (m, 3H), 7.31–7.28 (m, 2H), 7.21–7.16 (m, 5H), 7.13–7.10 (m, 3H), 7.07–7.05 (m, 2H), 7.03 (d, J = 14.3 Hz, 1H), 6.93–6.92 (m, 3H), 6.88–6.84 (m, 5H), 6.81–6.78 (m, 3H), 6.49 (d, J = 3.8 Hz, 1H).

^{13}C NMR (125 MHz, $CDCl_3$): δ 142.4, 141.4, 141.3, 141.1, 140.3, 140.0, 139.9, 139.7, 139.6, 139.3, 138.9, 132.9, 131.4, 131.3, 131.2, 130.8, 129.9, 127.9, 127.8, 127.5, 126.9, 126.6, 126.4, 126.4, 126.3, 125.7, 125.4, 125.2, 124.0, 122.3, 120.8, 120.3, 113.5, 110.5.

HRMS (APCI+): *m/z* calcd for $C_{48}H_{33}NS$ [M + H$^+$] 656.2406; found: 656.2403.

(*E*)-2,7-bis(4-isopropylphenyl)-9-(2-(5-phenylthiophen-2-yl)vinyl)-9*H*-carbazole (39):

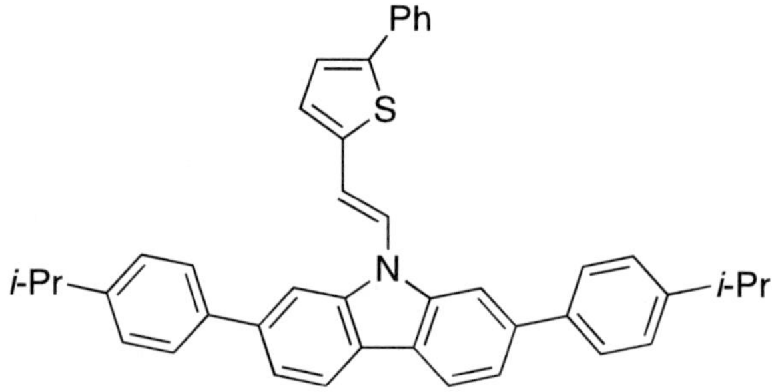

The title compound was obtained as yellow solid in 84% yield with *E*:*Z* = 7:1. The reaction was performed on a 0.20 mmol scale using 1.5 equiv of the thiophene, 1.0 equiv of the enamine, 20 mol % of $Fe(acac)_3$, and 22 mol % of TP.

Melting point: 230–234 °C (dichloromethane).

^{1}H NMR (500 MHz, $CDCl_3$): δ 8.12 (dd, J = 8.1, 0.6 Hz, 2H), 7.87 (d, J = 0.9 Hz, 2H), 7.71–7.64 (m, 7H), 7.56 (dd, J = 8.0, 1.4 Hz, 2H), 7.42–7.39 (m, 2H), 7.38–7.36 (m, 4H), 7.32–7.26 (m, 2H), 7.22 (dd, J = 14.3, 0.6 Hz, 1H), 7.06 (d, J = 3.7 Hz, 1H), 2.99 (hept, J = 6.9 Hz, 2H), 1.32 (d, J = 6.9 Hz, 12H).

^{13}C NMR (125 MHz, $CDCl_3$): δ 148.1, 142.3, 140.5, 139.8, 139.7, 139.3, 134.2, 129.0, 127.5, 126.9, 126.5, 125.6, 123.6, 122.9, 122.5, 120.7, 120.5, 114.3, 109.0, 33.8, 24.1. (one signal overlapped).

HRMS (APCI +): *m/z* calcd for $C_{42}H_{37}NS$ [M + H$^+$] 588.2719; found: 588.2731.

(*E*)-3,6-bis(4-methoxyphenyl)-9-(2-(5-phenylthiophen-2-yl)vinyl)-9*H*-carbazole (40):

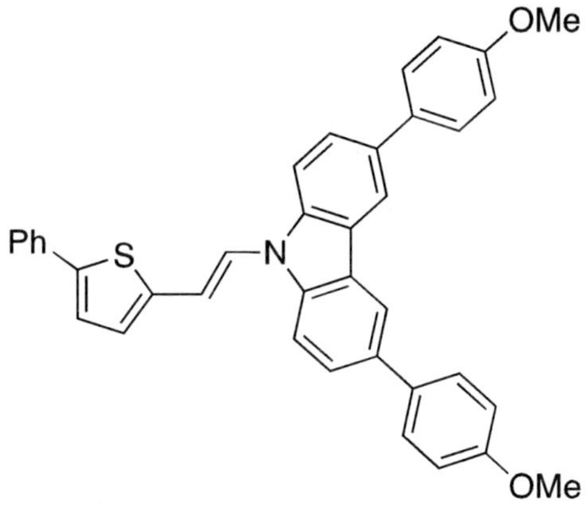

The title compound was obtained as orange solid in 76% yield with *E*:*Z* = > 20:1. The reaction was performed on a 0.20 mmol scale using 1.5 equiv of the thiophene, 1.0 equiv of the enamine, 20 mol % of $Fe(acac)_3$, and 22 mol % of TP.

Melting point: 192–193 °C (dichloromethane).

^{1}H NMR (500 MHz, $CDCl_3$): δ 8.28–8.27 (m, 2H), 7.76 (dd, J = 8.6, 0.6 Hz, 2H), 7.72–7.64 (m, 9H), 7.43–7.40 (m, 2H), 7.32–7.28 (m, 2H), 7.19 (dd, J = 14.4, 0.6 Hz, 1H), 7.07–7.03 (m, 5H), 3.89 (s, 6H).

^{13}C NMR (125 MHz, $CDCl_3$): δ 158.8, 142.1, 139.9, 138.8, 134.2, 134.1, 134.1, 129.0, 128.3, 127.5, 126.3, 125.6, 125.6, 124.8, 123.6, 122.6, 118.4, 114.3, 113.0, 110.9, 55.4.

HRMS (APCI +): *m/z* calcd for $C_{38}H_{29}NO_2S$ [M + H^+] 564.1992; found: 564.1985.

(*E*)-2-phenyl-1-(2-(5-phenylthiophen-2-yl)vinyl)-1*H*-indole (41):

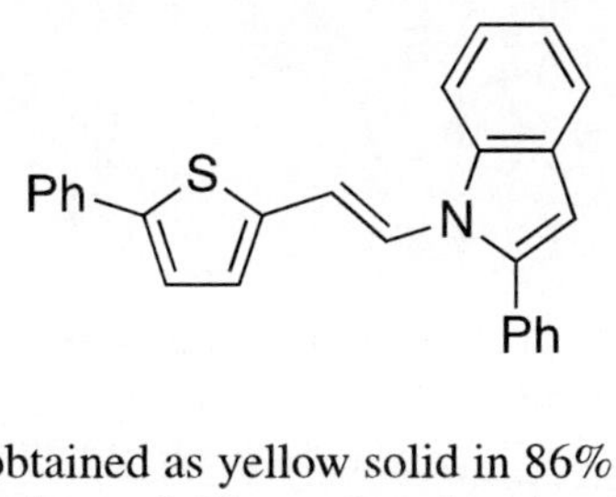

The title compound was obtained as yellow solid in 86% yield with *E*:*Z* = > 20:1. The reaction was performed on a 0.40 mmol scale using 1.5 equiv of the thiophene, 1.0 equiv of the enamine, 20 mol % of $Fe(acac)_3$, and 22 mol % of TP.

Melting point: 131–132 °C (dichloromethane).

^{1}H NMR (500 MHz, $CDCl_3$): δ 7.77–7.76 (m, 1H), 7.67–7.65 (m, 1H), 7.61–7.57 (m, 4H), 7.49–7.46 (m, 2H), 7.43–7.27 (m, 6H), 7.23–7.20 (m, 2H), 6.95–6.91 (m, 2H), 6.70 (s, 1H).

^{13}C NMR (125 MHz, $CDCl_3$): δ 142.4, 140.4, 139.4, 136.9, 134.0, 132.3, 129.4, 129.0, 128.9, 128.6, 128.1, 127.5, 126.5, 125.6, 124.4, 123.5, 122.8, 121.3, 120.8, 115.7, 111.4, 104.7.

HRMS (APCI +): *m/z* calcd for $C_{26}H_{19}NS$ [M + H$^+$] 378.1311; found: 378.1314.

(*E*)-2,3-diphenyl-1-(2-(5-phenylthiophen-2-yl)vinyl)-1*H*-indole (42):

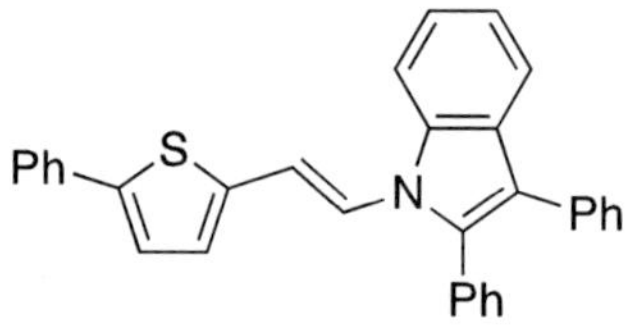

The title compound was obtained as yellow solid in 90% yield with *E*:*Z* = > 20:1. The reaction was performed on a 0.20 mmol scale using 1.5 equiv of the thiophene, 1.0 equiv of the enamine, 20 mol % of $Fe(acac)_3$, and 22 mol % of TP.

Melting point: 216–217 °C (dichloromethane).

^{1}H NMR (500 MHz, CD_2Cl_2): δ 7.76–7.74 (m, 1H), 7.65–7.63 (m, 1H), 7.50–7.48 (m, 2H), 7.32–7.14 (m, 17H), 6.84 (d, J = 3.8 Hz, 1H), 6.80 (dd, J = 14.3, 0.6 Hz, 1H).

^{13}C NMR (125 MHz, CD_2Cl_2): δ 142.4, 139.9, 136.8, 136.2, 134.6, 134.3, 131.7, 131.7, 130.3, 129.2, 128.8, 128.7, 128.7, 128.6, 127.9, 126.9, 126.5, 125.8, 124.4, 123.9, 123.5, 121.8, 120.0, 118.0, 115.4, 111.8.

HRMS (APCI +): *m/z* calcd for $C_{32}H_{23}NS$ [M + H$^+$] 454.1624; found: 454.1630.

(*E*)-2-phenyl-1-(2-(5-phenylthiophen-2-yl)vinyl)-1*H*-pyrrole (43):

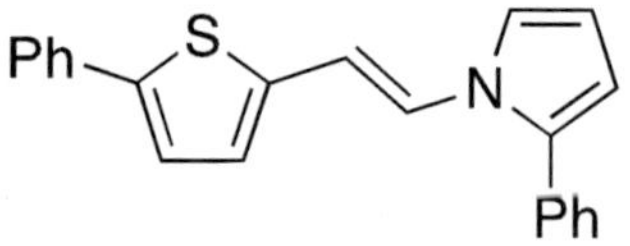

The title compound was obtained as yellow solid in 58% yield with *E*:*Z* = > 20:1. The reaction was performed on a 0.20 mmol scale using 1.5 equiv of the thiophene, 1.0 equiv of the enamine, 20 mol % of $Fe(acac)_3$, and 22 mol % of TP.

Melting point: 107–109 °C (dichloromethane).

^{1}H NMR (500 MHz, CD_2Cl_2): δ 7.49–7.47 (m, 2H), 7.38–7.35 (m, 4H), 7.30–7.26 (m, 3H), 7.23–7.17 (m, 2H), 7.14 (d, J = 3.7 Hz, 1H), 7.10 (dd, J = 2.9, 1.8 Hz, 1H), 6.88 (d, J = 3.7 Hz, 1H), 6.73 (d, J = 14.3 Hz, 1H), 6.27–6.23 (m, 2H).

^{13}C NMR (125 MHz, CD_2Cl_2): δ 142.2, 139.8, 134.8, 134.3, 132.6, 129.4, 129.2, 128.9, 127.8, 127.6, 126.8, 125.7, 125.6, 123.8, 119.2, 110.9, 110.6, 110.5.

HRMS (APCI +): *m/z* calcd for $C_{22}H_{17}NS$ [M + H$^+$] 328.1154; found: 328.1152.

(*E*)-*N*-phenyl-*N*-(2-(5-phenylthiophen-2-yl)vinyl)aniline (44):

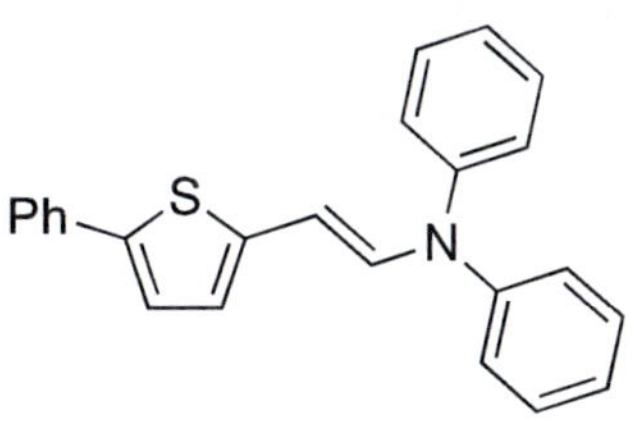

The title compound was obtained as yellow solid in 39% yield with E:$Z \geq$ 20:1. The reaction was performed at 70 °C on a 0.40 mmol scale using 1.5 equiv of the thiophene, 1.0 equiv of the enamine, 20 mol % of Fe(acac)$_3$, and 22 mol % of TP. Because of the instability of the product under acidic conditions, gel permeation chromatography (toluene) was used instead of silica gel chromatography for purification.

Melting point: 110–112 °C (toluene).

^{1}H NMR (500 MHz, DMSO-d_6): δ 7.09–7.07 (m, 2H), 6.98 (d, J = 13.7 Hz, 1H), 6.95–6.91 (m, 4H), 6.89–6.86 (m, 2H), 6.83 (d, J = 3.9 Hz, 1H), 6.75–6.68 (m, 3H), 6.61–6.59 (m, 4H), 6.32 (d, J = 3.9 Hz, 1H), 5.02 (d, J = 13.7 Hz, 1H).

^{13}C NMR (125 MHz, DMSO-d_6): δ 144.13, 142.0, 137.6, 133.9, 133.3, 129.9, 129.1, 126.9, 124.7, 124.4, 124.2, 123.3, 123.2, 102.0.

HRMS (APCI +): *m/z* calcd for $C_{24}H_{19}NS$ [M + H$^+$] 354.1311; found: 354.1307.

A general procedure for copolymerization (Table 5.6)

In an oven-dried Schlenk tube was added a thiophene monomer (0.10 mmol), an enamine monomer (0.10 mmol), TP (14 mg, 0.022 mmol), and a THF solution of Fe(acac)$_3$ (0.067 mol/L, 0.30 mL, 0.020 mmol). Then, a toluene solution of $AlMe_3$ (2.0 mol/L, 0.30 mL, 0.60 mmol) was added by rinsing the wall of the Schlenk tube. Diethyl oxalate (54 μ L, 0.40 mmol) was added, and the reaction mixture was stirred at 50 °C for 24 h. The reaction mixture was cooled to room temperature, diluted with chloroform (5 mL), quenched carefully with an aqueous solution of HCl (4 mol/L, 1.0 mL), and stirred for 10 min. The crude solution was diluted with chloroform (10 mL) and washed with water (10 mL × 2). The organic phase was dried over Na_2SO_4 and passed through a pad of Florisil. The solvent was removed under reduced pressure, and the crude product was purified by gel permeation chromatography (chloroform). After the removal of the solvent, the product was redissolved in a minimal amount of chloroform and precipitated from methanol to afford the final polymer product. *E*/*Z* ratio was determined by integration of the ^{1}H NMR spectrum. A chloroform solution of the product was prepared and analyzed by GPC to determine the M_n, M_w, M_w/M_n values of the isolated polymer. Degree of polymerization (DP) was determined by dividing the M_n value by the average molecular weight of two monomer units. Copolymers contained up to ~10% less an enamine unit than a thiophene unit as judged by integration of the carbazole protons in ^{1}H NMR spectra.

Poly(2,2'-(9,9-dioctyl-9*H*-fluorene-2,7-diyl)bis(3-hexylthiophene)-*co*-9,9'-divinyl-9*H*,9'*H*-3,3'-bicarbazole) (45)

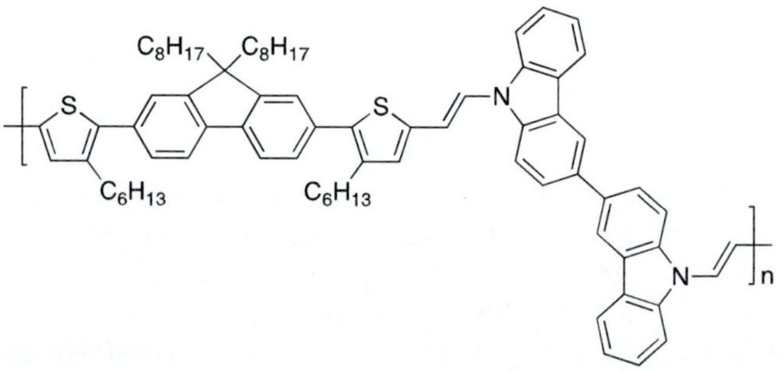

The title compound was obtained as yellow solid (98 mg, 89%, *E*:*Z* = 12:1). 2,2'-(9,9-Dioctyl-9*H*-fluorene-2,7-diyl)bis(3-hexylthiophene) (72 mg, 0.10 mmol) and 9,9'-divinyl-9*H*,9'*H*-3,3'-bicarbazole (38 mg, 0.10 mmol) were used as monomers. ^{13}C NMR spectrum could not be analyzed because of the weak intensity of the signals.

^{1}H NMR (500 MHz, $CDCl_3$, 55 °C): δ 8.44–8.37 (br, 2H), 8.22–8.15 (br, 2H), 7.90–7.65 (br, 8H), 7.58–7.31 (br, 10H), 7.24–7.00 (br, 4H), 2.75–2.72 (br, 4H), 2.06–2.01 (br, 4H), 1.72–1.68 (br, 4H), 1.44–1.10 (br, 32H), 0.90–0.77 (br, 16H).

$M_n = 13.1$ kg/mol, $M_w = 28.5$ kg/mol, $M_w/M_n = 2.18$, DP = 24.

Poly(*N*,*N*-bis(4-(3-hexylthiophen-2-yl)phenyl)-2,4,6-trimethylaniline-*co*-9,9'-divinyl-9*H*,9'*H*-3,3'-bicarbazole) (46):

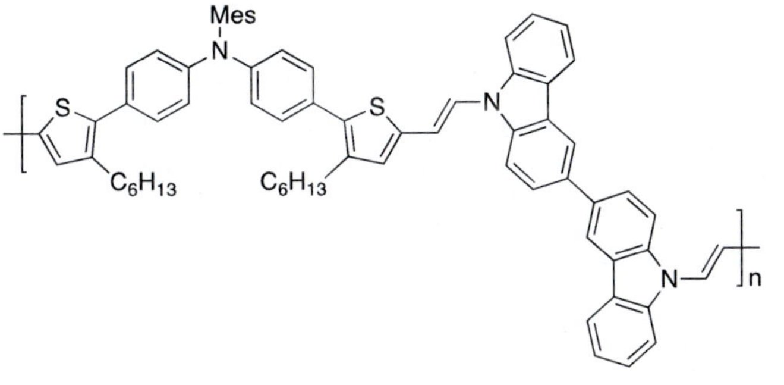

The title compound was obtained as yellow solid (87 mg, 92%, *E*:*Z* = 4:1). *N*,*N*-bis(4-(3-hexylthiophen-2-yl)phenyl)-2,4,6-trimethylaniline (62 mg, 0.10 mmol) and 9,9'-divinyl-9*H*,9'*H*-3,3'-bicarbazole (38 mg, 0.10 mmol) were used as monomers. ^{13}C NMR spectrum could not be analyzed because of the weak intensity of the signals.

^{1}H NMR (500 MHz, $CDCl_3$, 55 °C): δ 8.41–8.33 (br, 2H), 8.21–8.10 (br, 2H), 7.86–7.58 (br, 6H), 7.53–7.45 (br, 2H), 7.42–7.26 (br, 8H), 7.18–7.12 (br, 2H), 7.09–7.01 (br, 4H), 7.00–6.94 (br, 4H), 2.69–2.65 (br, 4H), 2.35 (br s, 3H), 2.09 (br s, 6H), 1.67–1.61 (br, 4H), 1.42–1.25 (br, 12H), 0.90–0.81 (br, 6H).

$M_n = 8.9$ kg/mol, $M_w = 18.6$ kg/mol, $M_w/M_n = 2.11$, DP = 19.

Poly(2-decyltetradecyl 3,5-bis(3-hexylthiophen-2-yl)benzoate-*co*-9,9'-divinyl-9*H*,9'*H*-3,3'-bicarbazole) (47):

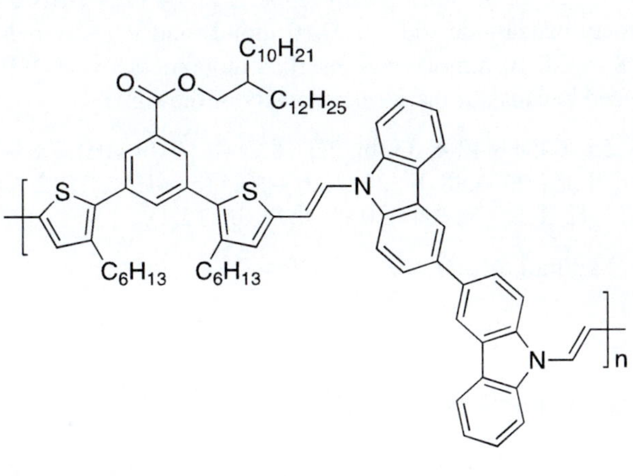

The title compound was obtained as yellow solid (101 mg, 86%, *E*:*Z* = 4:1). 2-Decyltetradecyl 3,5-bis(3-hexylthiophen-2-yl)benzoate (79 mg, 0.10 mmol) and 9,9'-divinyl-9*H*,9'*H*-3,3'-bicarbazole (38 mg, 0.10 mmol) were used as monomers. ^{13}C NMR spectrum could not be analyzed because of the weak intensity of the signals.

^{1}H NMR (500 MHz, $CDCl_3$, 55 °C): δ 8.42–8.30 (br, 2H), 8.20–8.06 (br, 4H), 7.89–7.61 (br, 9H), 7.54–7.46 (br, 2H), 7.36–7.26 (br, 2H), 7.23–7.14 (br, 2H), 7.05–6.99 (br, 2H), 4.31–4.29 (br, 2H), 2.76–2.68 (br, 4H), 1.84–1.80 (br, 1H), 1.78–1.62 (br, 4H), 1.35–1.17 (br, 52H), 0.89–0.80 (br, 12H).

$M_n = 11.0$ kg/mol, $M_w = 23.2$ kg/mol, $M_w/M_n = 2.11$, DP = 19.

Poly(3,5-bis(3-hexylthiophen-2-yl)-*N*,*N*-dioctylbenzamide-*co*-9,9'-divinyl-9*H*,9'*H*-3,3'-bicarbazole) (48):

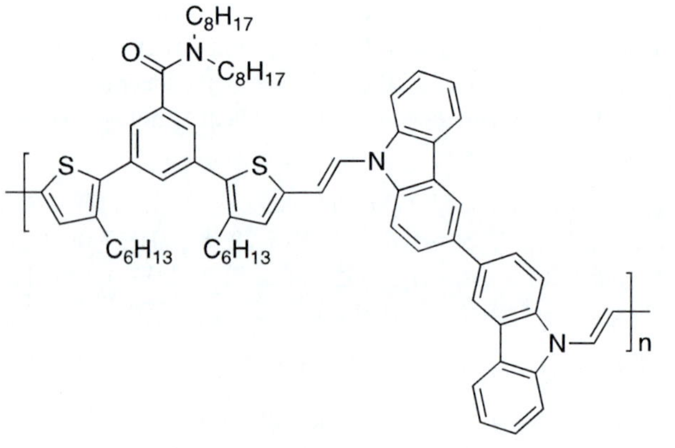

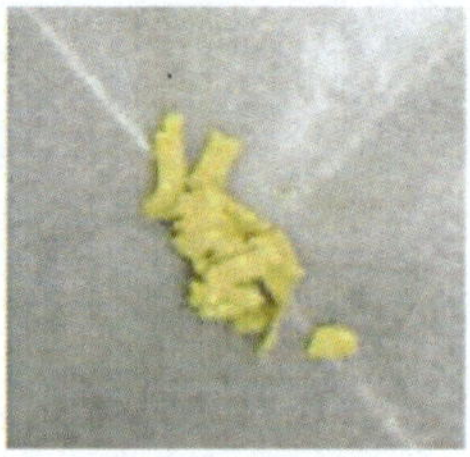

The title compound was obtained as yellow solid (92 mg, 87%, *E*:*Z* = 4:1). 3,5-Bis(3-hexylthiophen-2-yl)-*N*,*N*-dioctylbenzamide (68 mg, 0.10 mmol) and 9,9'-divinyl-9*H*,9'*H*-3,3'-bicarbazole (38 mg, 0.10 mmol) were used as monomers. ^{13}C NMR spectrum could not be analyzed because of the weak intensity of the signals.

^{1}H NMR (500 MHz, $CDCl_3$, 55 °C): δ 8.42–8.33 (br, 2H), 8.21–8.10 (br, 2H), 7.88–7.27 (br, 15H), 7.23–7.13 (br, 2H), 7.09–6.98 (br, 2H), 3.63–3.25 (br, 4H), 2.75–2.70 (br, 4H), 1.74–1.51 (br, 8H), 1.42–1.10 (br, 32H), 0.95–0.76 (br, 12H).

$M_n = 8.3$ kg/mol, $M_w = 16.2$ kg/mol, $M_w/M_n = 1.95$, DP = 16.

Poly(4,8-bis((2-octyldodecyl)oxy)benzo[1,2-*b*:4,5-*b'*]dithiophene-*co*-9,9'-divinyl-9*H*,9'*H*-3,3'-bicarbazole) (49):

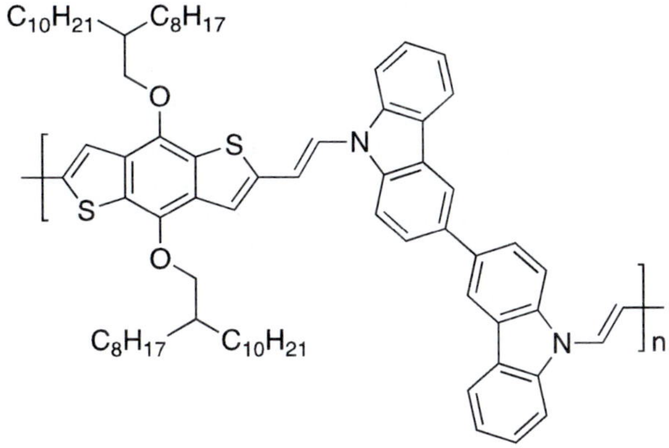

The title compound was obtained as dark orange solid (104 mg, 89%, E:Z = 19:1). 4,8-Bis((2-octyldodecyl)oxy)benzo[1,2-*b*:4,5-*b'*]dithiophene (78 mg, 0.10 mmol) and 9,9'-divinyl-9*H*,9'*H*-3,3'-bicarbazole (38 mg, 0.10 mmol) were used as monomers. ^{13}C NMR spectrum could not be analyzed because of the weak intensity of the signals.

^{1}H NMR (500 MHz, $CDCl_3$, 55 °C): δ 8.43–8.33 (br, 2H), 8.23–8.10 (br, 2H), 7.91–7.60 (br, 8H), 7.58–7.51 (br, 2H), 7.43–7.26 (br, 6H), 4.30–4.21 (br, 4H), 1.93–2.00 (br, 2H), 1.75–1.68 (br, 4H), 1.55–1.20 (br, 60H), 0.89–0.81 (br, 12H).

M_n = 10.6 kg/mol, M_w = 20.8 kg/mol, M_w/M_n = 1.97, DP = 18.

Poly(2,2'-(9,9-dioctyl-9*H*-fluorene-2,7-diyl)bis(3-hexylthiophene)-*co*-5,11-divinyl-5,11-dihydroindolo[3,2-*b*]carbazole) (50):

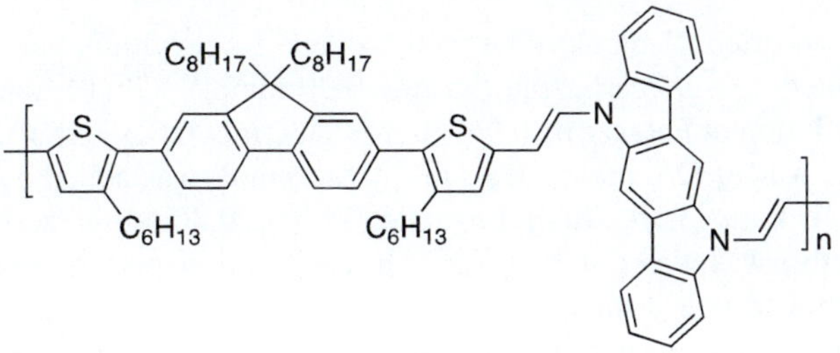

The title compound was obtained as yellow solid (78 mg, 76%, E:Z = 4:1). 2,2'-(9,9-dioctyl-9*H*-fluorene-2,7-diyl)bis(3-hexylthiophene) (72 mg, 0.10 mmol) and 5,11-divinyl-5,11-dihydroindolo[3,2-*b*]carbazole (31 mg, 0.10 mmol) were used as monomers. ^{13}C NMR spectrum could not be analyzed because of the weak intensity of the signals.

^{1}H NMR (500 MHz, $CDCl_3$): δ 8.36–8.79 (br, 4H), 7.77–7.70 (br, 4H), 7.60–7.26 (br, 10H), 7.15–6.83 (br, 4H), 2.78–2.66 (br, 4H), 2.07–1.97 (br, 4H), 1.76–1.64 (br, 4H), 1.43–1.04 (br, 32H), 0.93–0.72 (br, 16H).

$M_n = 5.9$ kg/mol, $M_w = 12.0$ kg/mol, $M_w/M_n = 2.03$, DP = 12.

Procedure for *EZ* isomerization (Scheme 5.10)

A mixture of (*E*)-9-(2-(5-phenylthiophen-2-yl)vinyl)-9*H*-carbazole (7.0 mg, 0.020 mmol) in THF (0.50 mL) was irradiated with purple LED (390 nm, 52 W) at rt for 4 h. The crude reaction mixture was passed through a pad of silica gel using ethyl acetate as an eluent, and the solvent was removed under reduced pressure to afford the product as pale yellow solid (6.9 mg, 99% yield, *E*:*Z* = 1:14).

(Z)-9-(2-(5-phenylthiophen-2-yl)vinyl)-9*H*-carbazole (3-Z):

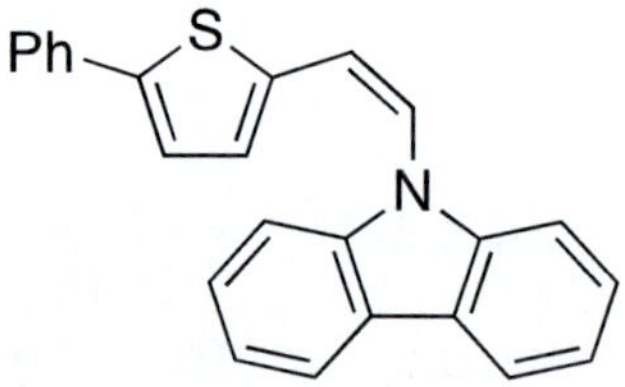

Melting point: 143–146 °C (ethyl acetate).

^{1}H NMR (500 MHz, $CDCl_3$): δ 8.14–8.12 (m, 2H), 7.42–7.39 (m, 2H), 7.32–7.27 (m, 6H), 7.24–7.21 (m, 2H), 7.18–7.15 (m, 1H), 7.06 (d, $J = 4.0$ Hz, 1H), 6.98 (d, $J = 8.2$ Hz, 1H), 6.89 (d, $J = 4.0$ Hz, 1H), 6.72 (d, $J = 8.2$ Hz, 1H).

^{13}C NMR (125 MHz, $CDCl_3$): δ 145.8, 139.6, 136.0, 133.8, 130.1, 128.7, 127.6, 126.0, 125.7, 124.2, 123.3, 122.6, 120.4, 120.3, 119.8, 110.9.

HRMS (APCI+): *m/z* calcd for $C_{24}H_{17}NS$ [M + H^+] 352.1154; found: 352.1160.

Procedure for deuterium labeling experiments (Scheme 5.11)

Scheme 5.11a: In an oven-dried Schlenk tube was added 2-phenylthiophene (32 mg, 0.20 mmol), 9-(vinyl-*d*3)-9*H*-carbazole (39 mg, 0.20 mmol), TP (14 mg, 0.022 mmol), and a THF solution of $Fe(acac)_3$ (0.067 mol/L, 0.30 mL, 0.020 mmol). Then, a toluene solution of $AlMe_3$ (2.0 mol/L, 0.30 mL, 0.60 mmol) was added by rinsing the wall of the Schlenk tube at rt. Diethyl oxalate (54 μ L, 0.40 mmol) was added, and the reaction mixture was stirred at 70 °C for 15 h. The reaction mixture was cooled to rt, diluted with ethyl acetate (1 mL), and quenched carefully with methanol (0.1 mL) and a saturated aqueous solution of potassium sodium tartrate (1 mL). The reaction mixture was diluted with toluene until all the products are dissolved and was stirred vigorously until clear phase separation was observed. Tridecane (30 μ L) was added as an internal standard and a portion of the organic layer was passed through a pad of Florisil and analyzed by GC to determine the yields. Then, the products were isolated by silica gel chromatography (hexane/dichloromethane = 19/1 to 9/1 to 4/1) and deuterium incorporation ratios were determined by ^{1}H NMR.

Scheme 5.11b: In an oven-dried Schlenk tube was added 2-phenylthiophene-5-*d* (32 mg, 0.20 mmol), 9-vinyl-9*H*-carbazole (39 mg, 0.20 mmol), TP (14 mg, 0.022 mmol), and a THF solution of $Fe(acac)_3$ (0.067 mol/L, 0.30 mL, 0.020 mmol). Then, a toluene solution of $AlMe_3$ (2.0 mol/L, 0.30 mL, 0.60 mmol) was added by rinsing the wall of the Schlenk tube at rt. Diethyl oxalate (54 μ L, 0.40 mmol) was added, and the reaction mixture was stirred at 70 °C for 15 h. The reaction mixture was cooled to rt, diluted with ethyl acetate (1 mL), and quenched carefully with methanol (0.1 mL) and a saturated aqueous solution of potassium sodium tartrate (1 mL). The reaction mixture was diluted with toluene until all the products are dissolved and was stirred vigorously until clear phase separation was observed. Tridecane (30 μ L) was added as an internal standard, and a portion of the organic layer was passed through a pad of Florisil and analyzed by GC to determine the yields. Then, the products were isolated by silica gel chromatography (hexane/dichloromethane = 19/1 to 9/1 to 4/1) and deuterium incorporation ratios were determined by ^{1}H NMR.

Procedure for kinetic isotope effect experiments (Scheme 5.12)

In an oven-dried Schlenk tube was added 2-phenylthiophene (32 mg, 0.20 mmol), 9-vinyl-9*H*-carbazole (39 mg, 0.20 mmol), TP (14 mg, 0.022 mmol), and a THF solution of $Fe(acac)_3$ (0.067 mol/L, 0.30 mL, 0.020 mmol). Then, a toluene solution of $AlMe_3$ (2.0 mol/L, 0.30 mL, 0.60 mmol) was added by rinsing the wall of the Schlenk tube at rt. Diethyl oxalate (54 μ L, 0.40 mmol) was added, the reaction mixture was stirred at 50 °C, and 0.1 mL of the reaction mixture was sampled after 0.5, 1, 2, 3, 4 h. Exactly the same experiment using the same solutions of reagents and the same reactor was conducted at the same time using 2-phenylthiophene-5-*d* (32 mg, 0.20 mmol) and 9-(vinyl-*d*3)-9*H*-carbazole (39 mg, 0.20 mmol). Each sample was quenched with a saturated aqueous solution of potassium sodium tartrate (0.5 mL), and the aqueous layer was extracted with ethyl acetate/toluene mixture (1:1, 0.5 mL). The organic layer was passed through a pad of Florisil and analyzed by GC using tridecane as an internal standard.

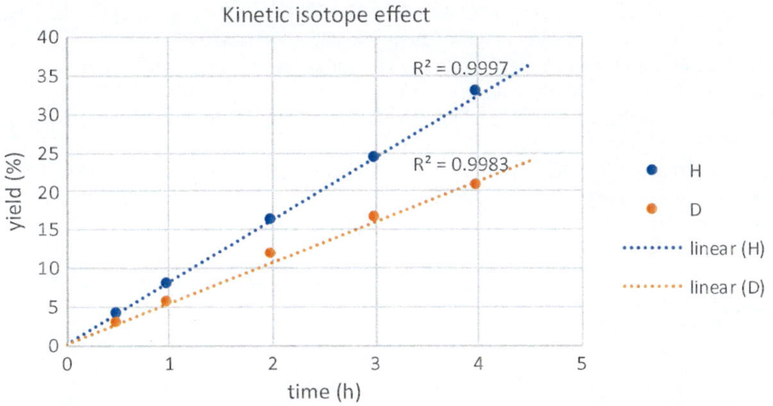

References

1. Moritani I, Fujiwara Y (1967) Tetrahedron Lett 8:1119–1122
2. Fujiwara Y, Moritani I, Danno S, Asano R, Teranishi S (1969) J Am Chem Soc 91:7166–7169
3. Gorelsky SI, Lapointe D, Fagnou K (2008) J Am Chem Soc 130:10848–10849
4. Murai S, Kakiuchi F, Sekine S, Tanaka Y, Kamatani A, Sonoda M, Chatani N (1993) Nature 366:529–531
5. Boele MDK, van Strijdonck GPF, de Vries AHM, Kamer PCJ, de Vries JG, van Leeuwen PWNM (2002) J Am Chem Soc 124:1586–1587
6. For reviews see: (a) Yeung CS, Dong VM (2011) Chem Rev 111:1215–1292 (b) Ma W, Gandeepan P, Li J, Ackermann L (2017) Org Chem Front 4:1435–1467 (c) Ali W, Prakash G, Maiti D (2021) Chem Sci 12:2735–2759
7. (a) Leow D, Li G, Mei T-S, Yu J-Q (2012) Nature 486:518–522 (b) Yang Y-F, Cheng G-J, Liu P, Leow D, Sun T-Y, Chen P, Zhang X, Yu J-Q, Wu YD, Houk KN (2014) J Am Chem Soc 136:344–355 (c) Dai H-X, Li G, Zhang X-G, Stepan AF, Yu J-Q (2013) J Am Chem Soc 135:7567–7571 (d) Tang R-Y, Li G, Yu J-Q (2014) Nature 507:215–220 (e) Yang G, Lindovska P, Zhu D, Kim J, Wang P, Tang R-Y, Movassaghi M, Yu J-Q (2014) J Am Chem Soc 136:10807–10813 (f) Deng Y, Yu J-Q (2015) Angew Chem Int Ed 54:888–891 (g) Chu L, Shang M, Tanaka K, Chen Q, Pissarnitski N, Streckfuss E, Yu J-Q (2015) ACS Cent Sci 1:394–399
8. Reports from other groups: (a) Lee S, Lee H, Tan KL (2013) J Am Chem Soc 135:18778–18781 (b) Bera M, Modak A, Patra T, Maji A, Maiti D (2014) Org Lett 16:5760–5763 (c) Bera M, Maji A, Sahoo SK, Maiti D (2015) Angew Chem Int Ed 54:8515–8519 (d) Li S, Ji H, Cai L, Li G (2015) Chem Sci 6:5595–5600 (e) Patra T, Watile R, Agasti S, Naveen T, Maiti D (2016) Chem Commun 52:2027–2030 (f) Li S, Cai L, Ji H, Yang L, Li G (2016) Nat Commun 7:10443
9. (a) Bag S, Patra T, Modak A, Deb A, Maity S, Dutta U, Dey A, Kancherla R, Maji A, Hazra A, Bera M, Maiti D (2015) J Am Chem Soc 137:11888–11891 (b) Patra T, Bag S, Kancherla R, Mondal A, Dey A, Pimparkar S, Agasti S, Modak A, Maiti D (2016) Angew Chem Int Ed 55:7751–7755
10. Wang P, Verma P, Xia G, Shi J, Qiao JX, Tao S, Cheng PTW, Poss MA, Farmer ME, Yeung K-S, Yu J-Q (2017) Nature 551:489–493
11. Reports from other groups: (a) Chen H, Wedi P, Meyer T, Tavakoli G, van Gemmeren M (2018) Angew Chem Int Ed 57:2497–2501 (b) Yin B, Fu M, Wang L, Liu J, Zhu Q (2020) Chem Commun 56:3293–3296
12. (a) Liu S, Saidi O, Berry N, Ruan J, Pettman A, Thomson N, Xiao J (2009) Lett Org Chem 6:60−64 (b) von Schenck H, Åkermark B, Svensson M (2002) Organometallics 21:2248–2253 (c) von Schenck H, Åkermark B, Svensson M (2003) J Am Chem Soc 125:3503–3508
13. Deb A, Maiti D (2017) Eur J Org Chem 2017:1239–1252
14. (a) Zhang H-J, Lin W, Su F, Wen T-B (2016) Org Lett 18:6356–6359 (b) Li Y, Xue D, Lu W, Fan X, Wang C, Xiao J (2013) RSC Adv 3:11463–11466 (c) Pankajakshan S, Xu Y-H, Cheng JK, Low MT, Loh T-P (2012) Angew Chem Int Ed 51:5701–5705
15. Lin W, Li W, Lu D, Su F, Wen T-B, Zhang H-J (2018) ACS Catal 8:8070–8076
16. (a) Tanaka M, Takada Y, Tamano M (2010) Jpn. Kokai Tokkyo Koho JP 2010126571 A, June 10, 2010 (b) Machida S, Iijima H, Nakada M, Hiraide M (2019) Jpn. Kokai Tokkyo Koho JP 2019169693 A October 3, 2019
17. (a) Pivsa-Art S, Satoh T, Kawamura Y, Miura M, Nomura M (1998) Bull Chem Soc Jpn 71:467–473 (b) Lane BS, Brown MA, Sames D (2005) J Am Chem Soc 127:8050–8057 (c) Würtz S, Rakshit S, Neumann JJ, Dröge T, Glorius F (2008) Angew Chem Int Ed 47:7230–7233 (d) Neumann JJ, Rakshit S, Dröge T, Würtz S, Glorius F (2011) Chem Eur J 17:7298–7303 (e) Shi Z, Glorius F (2012) Angew Chem Int Ed 51:9220–9222 (f) Ge H, Niphakis MJ, Georg GI (2008) J Am Chem Soc 130: 3708–3709
18. Lin H-S, Doba T, Sato W, Matsuo Y, Shang R, Nakamura E (2022) Angew Chem Int Ed 61:e202203949
19. Shen K, Fu Y, Li J-N, Liu L, Guo Q-X (2007) Tetrahedron 63:1568–1576

20. (a) Monks BM, Fruchey ER, Cook SP (2014) Angew Chem Int Ed 53:11065–11069 (b) Yoshikai N, Asako S, Yamakawa T, Ilies L, Nakamura E (2011) Chem Asian J 6:3059–3065 (c) Gu Q, Al Mamari HH, Graczyk K, Diers E, Ackermann L (2014) Angew Chem Int Ed 53:3868–3871
21. Simmons EM, Hartwig JF (2012) Angew Chem Int Ed 51:3066–3072
22. Gómez-Gallego M, Sierra MA (2011) Chem Rev 111:4857–4963
23. Still WC, Kahn M, Mitra A (1978) J Org Chem 43:2923–2925
24. Pangborn AB, Giardello MA, Grubbs RH, Rosen RK, Timmers FJ (1996) Organometallics 15:1518–1520
25. Doba T, Ilies L, Sato W, Shang R, Nakamura E (2021) Nat Catal 4:631–638
26. Tlahuext-Aca A, Hartwig JF (2021) ACS Catal 11:1119–1127
27. Ledovskaya MS, Voronin VV, Rodygin KS, Posvyatenko AV, Egorova KS, Ananikov VP (2019) Synthesis 51:3001–3013

Chapter 6
Conclusions and Perspectives

Transition-metal-catalyzed C(sp^2)–H/C(sp^2)–H coupling has attracted much attention as one of the most straightforward methods to construct C(sp^2)–C(sp^2) bonds. However, the application of this ideal transformation to the synthesis of redox-sensitive π-materials was hindered by the requirement of a strong oxidant for catalyst turnover. To overcome this problem, I focused on the low redox potential of Fe(III)/Fe(I) to achieve iron-catalyzed C–H/C–H coupling under mildly oxidative conditions.

Chapter 1 described that redox potential of catalyst is the limiting factor of the mildness of oxidant in transition-metal-catalyzed C–H/C–H coupling and proposed that the low redox potential of Fe(III)/Fe(I), which has been partially proved by the use of dihaloalkanes as a mild oxidant in iron-catalyzed C–H activation reactions, would be a solution to solve the problem.

Chapter 2 described the development of iron-catalyzed regioselective thienyl C–H/C–H homocoupling using tridentate phosphine as a ligand, $AlMe_3$ as a base, and oxalate as a mild oxidant. Tridentate phosphine ligand was uniquely effective for this transformation and oxalate served as an effective but mild oxidant to turnover the Fe(III)/Fe(I) catalytic cycle taking advantage of the oxophilicity of Al(III). The electronic bias created by a sulfur atom of thienyl groups helped to make the adjacent C–H bond acidic enough for iron-catalyzed C–H activation through a σ-bond metathesis mechanism. The reaction did not require a preinstalled directing group, giving direct access to useful π-conjugated dimeric and oligomeric thiophene compounds.

Chapter 3 described a modular method to synthesize conjugated tridentate phosphine ligands with different substituents. Sequential addition of two different organolithium reagents to $P(OPh)_3$ was the key to selectively obtain tridentate phosphine ligand out of bidentate or tetradentate ones. This method was applicable to the synthesis of aryl-TPs and heteroaryl-TPs, and the pure products were obtained in high yields on a gram scale by recrystallization. This method accelerated the further exploration of iron-catalyzed regioselective thienyl C–H/C–H coupling.

T. Doba, *Iron-Catalyzed C–H/C–H Coupling for Synthesis of Functional Small Molecules and Polymers*, Springer Theses,
https://doi.org/10.1007/978-981-99-4121-6_6

Chapter 4 described the development of iron-catalyzed regioselective thienyl C–H/C–H polycondensation by improvement of iron-catalyzed regioselective thienyl C–H/C–H homocoupling introduced in Chap. 2. Heteroaryl-TP was designed to suppress catalyst deactivation by prevention of intramolecular C–H activation of the ligand and effectively used as an optimal ligand for polycondensation. This reaction took place exclusively at the C–H bond next to the sulfur atom of thienyl groups, and no branching of the polymer was observed. Monomers containing various kinds of redox-sensitive π-motifs were polymerized efficiently up to DP of 46, which demonstrated the effectiveness and mildness of the iron catalytic cycle. Because of the weak interaction of Fe(III) with the π-surface of the polymer chain, polycondensation proceeded through a step-growth mechanism and the residual iron was removed to less than 40 ppm by post-treatment of a polymer with a thiol-functionalized silica scavenger. Moreover, a hybrid polymer of PTAA and P3HT was synthesized by this method and successfully applied as a hole-transporting material for perovskite solar cells in collaboration with Matsuo group. The device showed a highest PCE of 21.3% and a long-term stability over 1000 h, which could be explained by the perfect energy match toward $MAPbI_3$ and enhanced hydrophobicity by the hexyl substituents. This work highlighted the benefits of iron catalysis for the synthesis of π-conjugated polymeric compounds of importance in optoelectronic device applications.

Chapter 5 described the development of iron-catalyzed oxidative C–H alkenylation of thiophene compounds with enamines. The reaction gave direct access to donor materials containing unique vinylthiophene-carbazole connected structures with controlled regioselectivity, branched/linear selectivity, and *E*/*Z* selectivity. Owing to the high efficiency, this reaction was applicable to the synthesis of copolymers containing a bisthiophene and a bisenamine monomer units. Deuterium exchange between two substrates indicated that enamines react with thiophenes through an electrophilic metalation mechanism rather than a carbometallation mechanism.

In conclusion, iron-catalyzed C–H/C–H coupling reactions that proceed under mildly oxidative conditions were developed by utilizing the low redox potential of Fe(III)/Fe(I) cycle. Oxalate in combination with oxophilic Al(III) served as a mild oxidant to turn over the catalytic cycle and tridentate phosphine ligand was uniquely effective to control the reactivity of iron toward regioselective thienyl C–H activation. The reactions described herein highlighted the potential of iron, the most abundant transition-metal on earth, for the direct synthesis of functional small molecules and polymers of importance in materials science.

Finally, I would like to propose several directions for future research. First, the scope of this reaction system can be expanded to other (hetero)arenes by modulation of the ligand structure. Second, reactions of such (hetero)arenes with electrophiles are worth investigating. For example, carbonyl-containing electrophiles such as redox-active esters and *N*-benzoyloxyamines may accept electron(s) and serve as a coupling partner after fragmentation. Third, application of iron-catalyzed C–H/C–H coupling to the synthesis of π-materials that cannot be synthesized by other transition-metal-catalyzed reactions should be considered. Polymers containing unstable bonds such as C–B and B–N bonds would be the candidates. In addition, development of iron-catalyzed C–H activation reactions utilizing other C–H activation modes such as concerted-metalation-deprotonation (CMD) or oxidative addition would draw out the potential of iron to its fullest. Furthermore, a series of tridentate phosphine ligands that can be synthesized by the method described in Chap. 3 will give a great opportunity to investigate the catalysis of other transition metals. A compact tridentate coordination may be suitable for electron-deficient first-row transition metals with a small ionic radius.